Chemistry in the Soil Environment
ASA Special Publication Number 40

Proceedings of a symposium sponsored by Division S-2
of the American Society of Agronomy
and the Soil Science Society of America.
The papers were presented during the annual meetings
in Fort Collins, Colorado, August 5–10, 1979.

Organizing Committee
D. E. Baker

Editorial Committee
R. H. Dowdy
J. A. Ryan
V. V. Volk
D. E. Baker

Editor-in-Chief
Matthias Stelly

Managing Editor
David M. Kral

Assistant Editor
Mary Kay Cousin

1981
Published by the
AMERICAN SOCIETY OF AGRONOMY
SOIL SCIENCE SOCIETY OF AMERICA
677 South Segoe Road
Madison, Wisconsin 53711

American Society of Agronomy
Soil Science Society of America
677 South Segoe Road, Madison, Wisconsin 53711 USA

Library of Congress Catalog Card Number: 80-70243
Standard Book Number: 0-89118-065-6

Printed in the United States of America

FOREWORD

Managing chemicals in the environment is a major endeavor of man. The ever increasing number and complexity of synthetic chemicals as well as management of natural chemicals has been of tremendous benefit to humans. Fertilizers and pesticides are major tools in crop production but out of place, chemicals can be unsafe.

The soil is often a depository of chemicals whether intentionally or otherwise. Thus movement, sorption, and equilibria of chemicals in soils is fundamental to improved crop production and environmental management.

It is imperative that, in the years ahead, we develop ways to use chemicals more effectively for food and fiber production. At the same time, we must learn how to manage these chemicals in the soil so as to minimize health risks.

This publication is the result of a symposium held at the Soil Science Society of America meetings in August of 1979. The authors are experts in their field. The manuscripts, covering a wide variety of subjects, are an extremely valuable reference for the subject "Chemistry in the Soil Environment". The publication is in keeping with the Soil Science Societies goal of disseminating knowledge and technology for the improvement of man's total environment.

We are indebted to the authors, organizers, and editors for this fine volume.

R. L. Mitchell, ASA President, 1980
W. E. Larson, SSSA President, 1980

PREFACE

In recent years many soil chemists have emphasized research in a wide variety of applied soil's problems. Use of the basic soil chemistry principles helps elucidate and solve these problems. One manifestation of this trend has been the increased scope and variety of subject matter discussed within the Soil Chemistry Division (S-2) of the Soil Science Society of America as well as other divisions of this society and the American Society of Agronomy. Presentations of problem oriented research may not provide students and researchers who enter soil chemistry from related disciplines with the adequate perspective of soil chemical theory necessary for sustained research productivity. The elected program chairman for Division S-2 developed a program for the summer meetings held in Fort Collins, Colorado, in 1979, in which soil chemists presented their research in basic soil chemistry. Attendance of these sessions exceeded expectations. The desire of Division S-2 to publish the invited papers in a single volume led to this book.

Chemistry in the Soil Environment emphasizes the chemistry of elements accumulated by growing plants, their movement, sorption, and equilibria in soils, and the sources and types of charges driving these varied phenomena. The authors were chosen on the basis of their expertise in the given subject area. The authors were not asked to review all related literature, but were encouraged to integrate theory and concepts as they relate to soil properties and soil-plant-environment interrelationships. We are grateful for their contributions. They have provided us with a broad overview of important areas of applied and theoretical soil chemistry. The chapters in this volume will serve as "integrators of knowledge" worthy of emulation by aspiring soil chemists.

We wish to thank the many scientists who contributed suggestions and reviewed manuscripts during the preparation of this book.

Robert H. Dowdy, Chm., Editorial Committee
Dale E. Baker, Symposium Organizer
James A. Ryan, Committee Member
Van V. Volk, Committee Member

Contents

CHAPTER 1

Soil Chemistry and the Availability of Plant Nutrients[1]

STANLEY A. BARBER[2]

ABSTRACT

The plant availability of soil nutrients is determined by the nutrient supply characteristics of the soil and the nutrient absorption characteristics of the plant root. Soil supply is regulated by the rate of supply to the

---

[1] Journal paper no. 7836. Purdue Univ., Agric. Exp. Stn., West Lafayette, IN 47907. Contribution from the Dep. of Agronomy.

[2] Professor of agronomy.

root by mass-flow and diffusion. Nutrient uptake by the plant is influenced by the root system size and the rate of nutrient uptake per cm^2 of root as related to nutrient concentration at the root surface. A mathematical simulation model has been developed to describe the kinetics of nutrient uptake and it has been verified by P and K uptake by corn (*Zea mays* L.). The soil parameters in the model are: 1) concentration of the nutrient in the soil solution; 2) buffer power of the nutrient on the solid phase for replenishing the nutrient concentration in solution as it is depleted by plant absorption; and 3) the effective diffusion coefficient for the diffusion of the nutrient through the soil. The effect of soil chemical properties on these three parameters will influence the availability of soil nutrients to plants.

INTRODUCTION

Soil is the major nutrient source for plants growing in soil. Hence, the chemistry of these nutrients in the soil is significant in determining their availability for absorption by the plant root. Soil chemistry involves descriptive chemistry of the soil components, thermodynamics of the equilibrium relation between nutrients in various soil phases, and kinetics of the movement rate of nutrients between phases, such as occurs in soil near the plant root when nutrients are removed by absorption into the root. This treatise will discuss the chemistry I believe has direct major involvement in the immediate supply of nutrients from the soil to the plant root. I will not consider the long term effects of nutrient release from soil mineral weathering and microbiological action. For purposes of brevity, phosphate and K will be the only nutrients discussed. Available nutrients in the soil are defined as those nutrients that can reach the root and be absorbed at a rate sufficient to influence plant growth.

Methodology for measuring available plant nutrients in the soil has been under investigation for over 150 years (Melsted and Peck, 1973). Commonly, comparisons have been made of the correlation between each of several laboratory chemical extraction procedures and some plant measurement of nutrient availability. The plant measurement of nutrient availability may be the amount of nutrient absorbed by the plant, the nutrient composition of the plant, the increase in nutrient composition in the plant with addition of nutrient to the soil or the increase in plant growth resulting from nutrient addition to the soil. While satisfactory correlations may occur for studies on similar soils, they frequently are not satisfactory when a wide range of soil types is included in the study. Hanway (1973) reports,

———crop responses to fertilizer applications have exhibited a generally large variability not explained by the laboratory index of nutrient availability in the soil.

Progress in developing an understanding of the importance of soil chemistry in plant nutrient availability may be attained by studying the processes involved in the plant-root-soil system that determine the rate of flow of soil nutrients into the root. Nutrient flow through the soil and nutrient absorption into the root are dynamic processes so that kinetics may be a more important branch of chemistry than thermodynamics which concerns systems at equilibrium. The chemical properties of the plant root surface, the soil surfaces, and their interaction are important as we investigate the mechanisms that control the rate of flow of nutrients from the soil to the root and absorption into the root.

NUTRIENT SUPPLY MECHANISMS

Plant roots grow through the soil and absorb nutrients. Nutrients must be at the soil-root interface before they are positionally available for absorption into the root. A small amount of P and K will be at this interface when the root grows through the soil. Most of the P and K that the plant requires must move through the soil to the root before it becomes positionally available for uptake (Barber, 1962). Nutrients move to the root by mass-flow and diffusion. Mass-flow is the transport of nutrients in the convective flow of water to the root that results from plant transpiration. If mass-flow does not bring sufficient nutrient to the root system to meet the plant requirement, absorption by the root will reduce the P and K concentration in the soil at the soil-root interface and create a concentration gradient perpendicular to the root. Phosphate and K ions will diffuse along this gradient to the root due to the thermal motion of the ions. If mass-flow supplies more nutrients to the root than the root absorbs, the concentration in the soil at the soil-root interface will increase and nutrients will diffuse away from the root along the concentration gradient. The rate of nutrient absorption by the root will depend both on the rate of mass-flow and diffusive supply by the soil and the nature of the root absorption properties.

NUTRIENT ABSORPTION PROPERTIES OF THE ROOT

Rate of nutrient absorption per unit of root surface area is called influx, the balance between influx and nutrients lost by efflux is net influx. Net influx varies with species, varieties, temperature, plant age, root age, plant nutrient status, and nutrient concentration at the root surface. Since we are considering soil chemistry we can hold all of these variables constant except nutrient concentration at the root surface. Influx increases with increased concentration at a gradually decreasing rate of increase as shown by the curve between net influx and external concentration shown in Fig. 1. With increased external concentration, influx eventually reaches a maximum value called Imax. The curve has been described mathematically by the Michaelis-Menten equation, Eq. [1]

$$I = \frac{Imax\ C_l}{Km + C_l} \qquad [1]$$

where I is influx, C_l is nutrient concentration in a stired solution external

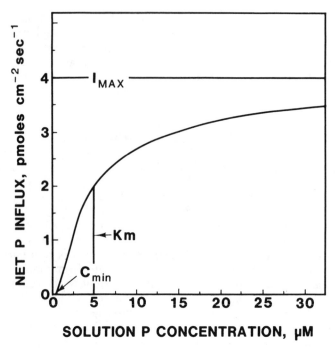

Fig. 1. The relation between influx of P and P concentration in the stirred solutions exterior to the roots.

to the root and Km is C_l where I is 1/2 Imax. Since plants do not completely deplete a solution, there is a value of C_l where In, net influx $= 0$, this is called Cmin.

MODELING NUTRIENT ABSORPTION FROM THE SOIL

Nutrient absorption by plants growing in soil represents a combination of the supply of nutrients from the soil by mass-flow and diffusion and absorption of the nutrients by the root at a rate dependent on the nutrient concentration maintained at the root surface. This process can be described by an objective mathematical simulation model. Claassen and Barber (1976) developed a model using radial coordinates to describe this process. They assumed that the roots were smooth cylinders that absorbed nutrients according to the relation shown in Fig. 1 and described by Eq. 1. The simulation model has 10 parameters. Three of the parameters describe the size of the root surface and its increase with time. Three parameters describe the relation between In and C_l illustrated in Fig. 1. One parameter is the rate of water absorption and three parameters describe the rate of nutrient supply to the root by the soil. The three soil parameters are:

1. The concentration of the nutrient in the soil solution before plant growth starts, C_{li}.

2. The buffer power, b, which measures the power of the concentration, C, of nutrients on the solid phase to buffer the concentration, C_l, of nutrients in the solution, as nutrients are removed from solution by root absorption. It is $\Delta C / \Delta C_l$, and may vary with C_l.

3. The effective diffusion coefficient, De, which describes the rate nutrients diffuse along a concentration gradient to the plant root. The rate of diffusion through the soil will be affected by soil chemistry.

MEASURING SOIL NUTRIENT SUPPLY PARAMETERS

The three parameters, C_{li}, b, and De which are used in the Claassen-Barber model to describe nutrient supply to the plant root can be measured in the laboratory. The value for C_{li} is obtained by displacing soil solution from the soil and measuring the nutrient concentration in the solution. Values for b are obtained by running a desorption or adsorption isotherm of C vs. C_l on the soil. The desorption isotherm is more relevant. However, since an adsorption isotherm is easier to determine, it can be used where the adsorption and desorption isotherms are similar as with K. For P the desorption isotherm is preferred since it may differ from the adsorption isotherm. The isotherms are usually conducted in 0.01 M CaCl$_2$ since this is about the ionic strength found in soil solution.

Values for De can be found by measuring diffusion in soil as described by Vaidyanathan and Nye (1966) or by calculation (Nye, 1968). Calculation involves use of Eq. [2]

$$De = D \, \theta \, \partial / b \qquad [2]$$

where D is the diffusion coefficient of the nutrient in water, θ is the volumetric water content, and ∂ is the tortuosity factor which accounts for the tortuous path the nutrient diffuses through the soil. In this equation, b has a very large effect in determining De for P and K. D can be obtained from chemical handbooks. ∂ is dependent on θ and soil texture (Warncke and Barber, 1972) and can be measured by measuring diffusion of a non-absorbed ion, such as Cl$^-$, through the soil.

VERIFICATION OF THE MODEL

Claassen and Barber (1976) verified the model for K uptake by growing corn (*Zea mays* L.) for 17 days in pots in a growth chamber. The comparison of K uptake predicted from the simulation model and observed uptake are shown in Fig. 2. There was overprediction of uptake which could have been due to the model not allowing for competition between roots for the K in the pot. The simulation model parameters for the soil were measured in the laboratory on soil that had the same previous treatment as that used for the plant uptake experiment. The K uptake parameters for corn were measured on the same age corn grown in solution culture. The root size and growth rate was measured from the experiment.

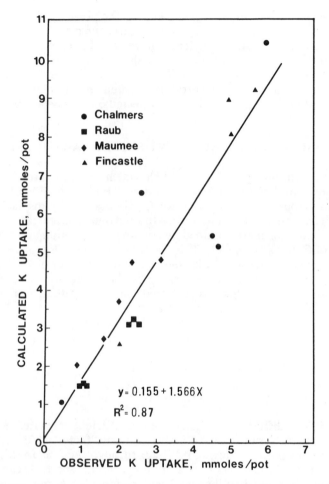

Fig. 2. Relation between observed K uptake by corn grown for 17 days in a controlled climate chamber and K uptake predicted by the Claassen-Barber model.

Schenk and Barber (1979) verified the model for P uptake by corn in pots in the growth chamber using five soils differing in P adsorption properties. The relation between predicted P uptake from the simulation model and observed P uptake is shown in Fig. 3. There was good agreement between predicted and observed uptake. There was little competition for P between adjacent roots because P does not move far in the soil. It has a much lower De than K. Verification for both P and K uptake indicates that in general, the model predicts P and K uptake and that the three soil parameters used are apparently the most important ones.

MEASURING SOIL NUTRIENT AVAILABILITY CHEMICALLY

The simulation model indicates that three soil parameters are important for describing the ability of the soil to supply nutrients to the

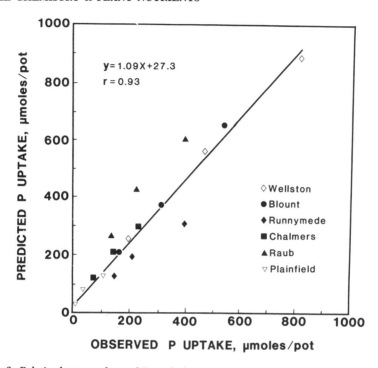

Fig. 3. Relation between observed P uptake by corn grown for 23 days in a controlled climate chamber and P uptake predicted by the Claassen-Barber model.

plant root. This makes measurement more complex than if only one parameter would suffice. Most soil test correlation studies have been attempts to use one or at most two parameters to obtain a soil nutrient availability index. From the investigations to date one could conclude that the only condition where one measurement, say C_{li}, would be a reliable index of soil nutrient availability would be where b and De were closely correlated with C_{li}. This may occur in soils that are very similar in all their chemical properties except for P levels.

Attempts have been made to use C_{li} and b as measures of nutrient availability (Fox and Kamprath, 1970). This procedure assumes a particular level of C_{li} is needed for the adequate nutrition of the crop and the measure of the P addition to reach this level is reflecting the magnitude of b. Commonly soil test procedures attempt to measure some index of C. This procedure is useful on similar soils where values of C are closely correlated with values for C_{li} and De on these soils. The use of intensity and capacity has also been an approach to evaluating soil nutrient availability. Intensity is usually described by measuring C_{li} while capacity is either measured with b or C. To get these values it is necessary to determine absorption curves (Beckett, 1964), however, simplifications have been proposed (Fox and Kamprath, 1970). The value of C_{li} divided by the square root of Ca plus Mg is frequently used as the abscessa and C on the ordinate when the data for a soil are plotted for evaluation.

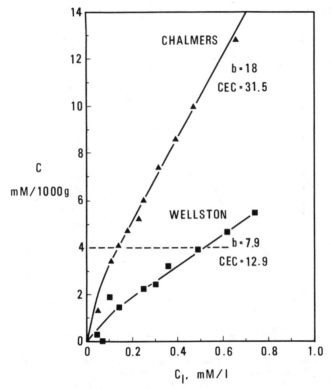

Fig. 4. Relation between K concentration in the soil solution and K concentration in the soil for Chalmers and Wellston soils.

An example of quantity-intensity curves that have been modified to give the important soil parameters is shown in Fig. 4. C_l for K instead of $K/\sqrt{Ca+Mg}$ is used on the abscissa. Instead of the change in exchangeable K based on the initial value in the soil we have shifted the base to show level of exchangeable K. Hence, if the ordinate were then changed to exchangeable K per unit volume by multiplying by the soil bulk density, we would have the values used for the buffer power isotherm. The slope of the curve then becomes b. Since we did not know the bulk density for the soils shown, we have assumed a bulk density value of 1.0 for purposes of illustration. The intensity-capacity relations that have been discussed (Beckett, 1964) usually measure values of C_{li} and b or indexes of them. These are two of the three parameters used in the Claassen-Barber simulation model. The third parameter, De, is greatly influenced by b, hence, for soils of similar water holding capacity and texture, De may be closely correlated with b, and b and C_{li} may be sufficient to describe nutrient availability on these soils.

The remainder of this treatise deals with the use of simulation models to explain various soil and plant root effects on soil nutrient availability.

Table 1. The C_{li}, b, C, and De values for P and K in the topsoil and subsoil of Raub silt loam.

Soil property	Phosphorus		Potassium	
	Topsoil	Subsoil	Topsoil	Subsoil
C_{li} μmoles/cm³ × 10²	1.1	0.02	51	15
C μmoles/cm³	4.23	1.23	5.75	3.08
b	400	6,150	11.3	20.5
De cm²sec × 10⁹	0.74	0.069	0.58	0.46

COMPARING P AND K UPTAKE FROM SURFACE AND SUBSOILS

The relative strengths of adsorption of P and K by soil surfaces may be different for surface soils then for subsoils. The values for C may not be correlated with values for C_{li} and b, hence, measuring only C may not give a satisfactory prediction of the available P or K in the soil. The C, C_{li}, b, and De values for P and K were measured on surface and subsoil samples immediately after collection without drying from Raub silt loam, Aquic argiudoll. The values are shown in Table 1. The P and K uptake from these soils was predicted with the simulation model since it has given close agreement between predicted and observed uptake in other experiments (Claassen and Barber, 1976; Schenk and Barber, 1979). Holding all root parameters constant, the predicted K uptake from the surface soil was 2.8 times that from the subsoil, whereas for K, C_{li} was 3.4, C was 1.9, and De was 1.25 times greater in the surface soil than in the subsoil. The predicted difference in K uptake between these soils represented the interaction of all parameters.

No P uptake was predicted from the subsoil because C_{li} for P was less than Cmin. Corn grown for 4 weeks on the subsoil contained less P than that present in the seed. The subsoil value of C was 29% of the surface soil value, whereas the subsoil C_{li} was only 2% of that in the surface soil. Therefore, single values would not be good predictors of P availability.

EFFECT OF CATION EXCHANGE CAPACITY, CEC, ON K AVAILABILITY

There has been a belief (McLean, 1976) that K availability is affected by CEC because with higher CEC soils, more Ca will be present and the proportion of K relative to Ca in the soil solution will be lower. The data of Fig. 4 lend themselves to an evaluation of this comparison using the simulation model. The pertinent data taken from these two soils are shown inTable 2 using a C value of 4 mmoles/1,000 g which is equivalent to 156 ppm. Uptake was simulated using a newer simulation model of Cushman (1979) which corrects for root to root competition. This model gives identical values to the Claassen-Barber where root competition does

Table 2. The C_{li}, b, and De values for K on Chalmers silty clay loam and Wellston silt loam where C is 4 mmoles/kg.

Soil property	Chalmers	Wellston
CEC me/100 g	31.5	12.9
C_{li} mM	0.14	0.52
C mmoles/kg	4	4
b	18	7.9
De cm²/sec × 10⁸	6.6	7.5
Uptake/week μmoles/plant	70	119

L_v, 3 cm cm⁻³; Imax, 41.5 pmoles cm⁻² sec⁻¹; Km, 20 μM; Cmin, 2.7 μM; Lo, 200 cm; k, 1 × 10⁻⁶ sec⁻¹, r, 0.02 cm.

Table 3. Values of C_{li}, b, De, and C required on Chalmers soil to give K uptake equal to K uptake from Wellston soil with same root growth.

Soil property	Chalmers	Wellston
C_{li}mM	0.215	0.52
C mmoles/kg	5.4	4
b	18	7.9
De cm²/sec × 10⁸	6.6	7.5
Uptake, μmoles/plant per week	119	119

not occur. A root density in the soil, L_v, of 3 cm per cm³ was assumed and uptake was for a 1-week period. Predicted uptake was less with the Chalmers soil having an exchange capacity 2.4 times greater.

The answer usually required from the data of Table 2 is what C value would be required in the Chalmers soil to give the same uptake as Wellston soil. The answer is given in Table 3 which shows that uptake was equal when C was 5.4 or about one-third higher. However, one should not conclude from this that C plus a modification for CEC will be a reliable predictor. Shaw (J. Shaw, 1979. Evaluation of differences in K availability in soils of the same exchangeable K level. M.S. thesis, Purdue Univ., W. Lafayette, Ind.) found that there were greater differences in K uptake between soils of the same CEC than between soils differing by 100% in CEC.

EFFECT OF ROOT DENSITY

The more roots present in the soil, the more competition there will be between roots and the greater the fraction of nutrients removed from the soil. A sensitivity analysis was conducted using the Cushman simulation model and two root densities and three soils which vary in CEC. The assumed data are given in Table 4 to illustrate how varying C_{li}, b, De, and root radius influence the K uptake predicted by the model. It was assumed that when CEC increased, C_{li} decreased for a constant C. This altered b which in turn altered De. The CEC for soil III is much higher than that for soil I. The root densities assumed were L_v = 40 cm/cm³ and 0.5 cm/cm³. This represents a relatively high root density and a rather low root density. Corn root density usually ranges from one to five in the surface soil for field grown corn (Mengel and Barber, 1974).

Table 4. Predicted K uptake as affected by interaction of root density and K absorption properties of the soil.

Soil property	Soil		
	I	II	III
C mmoles/kg	2	2	2
C_{li} mM	1	0.2	0.1
b	2	10	20
De, cm^2/sec $\times$ 10^7	5	1	0.5
Uptake μmoles/plant/week with			
$\quad r_1 = 0.09$ cm	23	21	20
$\quad r_1 = 0.45$ cm‡	269	121	71

The data show that where L_v is high and the mean one-half distance between roots is 0.09 cm, uptake is not greatly influenced by changing C_{li}, b, and De. The value of C is more important. Potassium uptake was 58 to 68% of C. Where L_v was 0.5 and the one-half distance between roots was 0.45 cm, there was much more soil per cm of root and much more K that could diffuse to the root. Since there was little root to root competition, changing C_{li} while holding C constant affected calculated uptake.

This shows that soils with high CEC may need higher C values for equal rates of soil supply per unit of root, but the difference rapidly becomes less as the root density increases. These data were extrapolations of observed parameters. The observations need experimental verification.

LITERATURE CITED

1. Barber, S. A. 1962. A diffusion and mass-flow concept of soil nutrient availability. Soil Sci. 93:39–49.
2. Beckett, P. H. T. 1964. Studies on soil potassium II. The immediate Q/I relations of labile potassium in the soil. J. Soil Sci. 15:9–23.
3. Claassen, N., and S. A. Barber. 1976. Simulation model for nutrient uptake from soil by a growing plant root system. Agron. J. 68:961–964.
4. Cushman, J. H. 1979. An analytical solution to solute transport near root surfaces for low initial concentrations. Soil Sci. Soc. Am. J. 43:1087–1095.
5. Fox, R. L., and E. J. Kamprath. 1970. Phosphate sorption isotherms for evaluating the phosphate requirements of soils. Soil Sci. Soc. Am. Proc. 34:902–906.
6. Hanway, J. J. 1973. Experimental methods for correlating and calibrating soil tests. In L. M. Walsh and J. D. Beaton (ed.) Soil testing and plant analysis. Rev. ed. Soil Sci. Soc. Am., Madison, Wis.
7. McLean, E. O. 1976. Exchangeable K levels for maximum crop yields on soils of different cation exchange capacities. Commun. Soil Sci. Pl. Anal. 7:823–838.
8. Melsted, S. W., and T. R. Peck. 1973. The principles of soil testing. In L. M. Walsh and J. D. Beaton (ed.) Soil testing and plant analysis. Rev. ed. Soil Sci. Soc. Am., Madison, Wis.
9. Mengel, D. B., and S. A. Barber. 1974. Development and distribution of corn root systems under field conditions. Agron. J. 66:341–344.
10. Nye, P. H. 1968. The use of exchange isotherms to determine diffusion coefficients in soil. Int. Soil Sci. Soc. Congr. Trans. 9th (Adelaide, Aust.) 1:117–126.
11. Schenk, M. K., and S. A. Barber. 1979. Phosphate uptake by corn as affected by soil characteristics and root morphology. Soil Sci. Soc. Am. J. 43:880–883.

12. Vaidyanathan, L. W., and P. H. Nye. 1966. The measurement and mechanism of ion diffusion in soil. II. An exchange resin paper method for measurement of the diffusive flux and diffusion coefficients of nutrient ions in soil. J. Soil Sci. 17:175–183.

13. Warncke, D. D., and S. A. Barber. 1972. Diffusion of zinc in soils. I. The influence of soil moisture. Soil Sci. Soc. Am. Proc. 36:39–42.

QUESTIONS AND ANSWERS

Q. What do you see as a short long-term resolution between quick soil test methods that require a rapid turnover time and measurement of buffer capacity, activities, in solution, and diffusion coefficients in soils?

A. The three parameters measuring soil supply in my paper are concentration of the nutrient in the soil solution, buffer power of the nutrient on the solid phase for the nutrient in solution, and apparent diffusion coefficient for the nutrient in the soil. These measurements are too time consuming for soil testing and are for research studies of the mechanisms. The amount predicted from the combined effect of these three parameters may correlate highly with a simple soil test and hence the soil test will be satisfactory for describing soil nutrient availability. Where the soil test does not correlate with nutrient availability for some soils in a particular area, measurement of the three parameters on these soils may tell you the reason. This may also enable you to predict the type of soil test that will give better correlation with plant availability. The measurements I have discussed are for use in soil fertility research to increase our understanding of the processes involved when nutrients are taken from soil by plant roots.

CHAPTER 2

Cation Exchange in Soils: An Historical and Theoretical Perspective[1]

GARRISON SPOSITO[2]

The historical development of the thermodynamic theory of cation exchange equilibria in soils is reviewed. The growth of this theory, a phenomenon of the present century, may be divided into three inter-related steps: 1) the evolution of the concept of an exchange equilibrium constant defined in terms of exchanger phase activities; 2) the unsuccessful search, through empirical postulates, for a universally-applicable expression for the exchange equilibrium constant in terms of composition variables; and 3) the ultimate derivation of a general thermodynamic equation for the exchange equilibrium constant in terms of an integral over a function of the equivalent fraction of an exchanger component. It

[1] Contribution from the Dep. of Soil and Environ. Sci., Univ. of California, Riverside 92521. Presented before Division S-2 of the Soil Sci. Soc. Am. at Fort Collins, Colo., on 8 Aug. 1979.
[2] Professor of soil science.

is concluded that these three steps may be interpreted as markers along the often irregular path to discover the most general expression for the dependence of exchanger activity coefficients on exchanger composition.

INTRODUCTION

Cation exchange in soil is a phenomenon known, at least through allusion, since the time of Aristotle (Forster, 1927, §933[b]). Its beginnings as a serious branch of chemistry, however, date from the 19th century in the work of Gazzari and of Way (Jenny, 1932; Kelley, 1948; Thomas, 1977). Much has been learned about soil exchange reactions since the results of the earliest research appeared in print and there is no doubt that the detailed study of these reactions has become the hallmark of modern soil chemistry. It is perhaps inevitable that a strong interest has arisen in recent years concerning the historical development of the experimental methods and the theoretical concepts pertaining to ion exchange in soils.

The motivation for this recent interest can be both pragmatic and aesthetic. On the one hand, an historical review of a scientific subject is the best method to avoid the continual rediscovery of an empirical fact, or conceptual model, and to understand the assumptions which underlie accepted theories. Thus, a careful developmental survey should result in more efficient scientific investigations. On the other hand, an historical review necessarily brings one into direct contact with the papers of those remarkable men and women who originated the important techniques and concepts that today appear routinely in textbooks. It is more often true than not that these seminal papers give the clearest, most carefully qualified, and most elegant expositions of the scientific results which since have become standard information. Perhaps it behooves every soil scientist with a concern for precision in understanding to read the words of these giants untranslated by pygmies.

Just 3 years ago, Thomas (1977) presented a lively and pertinent review of the development of knowledge concerning ion exchange reactions in soils. His survey covered the experimental side of cation exchange and provided a needed successor to the review given by Kelley (1948) some 30 years before. However, relatively little space in Thomas' (1977) paper was devoted to the historical development of the mathematical equations and other theoretical concepts which describe cation exchange in soils. The principal objective of this paper is to fill out some aspects of this area. Excellent critical reviews of cation exchange theory have been published by Helfferich (1962), Babcock (1963), Reichenberg (1966), Bolt (1967), and Reddy (1977). However, none of these reviews has considered the historical development of the theory in much detail.

As with any chemical phenomenon, cation exchange in soils has both a kinetics and an equilibrium aspect and may be approached from either the macroscopic or the molecular point of view. This paper will deal only with the equilibrium aspects (in part because they have been studied longer and are better understood) and will describe only the development of the macroscopic theory of cation exchange. This latter restriction of the topic will help to emphasize the fact that the macroscopic theory of exchange equilibria remains strictly independent of the molecular viewpoint, while at the same time providing the general description of empirical exchange data that a successful molecular theory is required to explain.

THE EXCHANGE EQUILIBRIUM CONSTANT

The macroscopic description of cation exchange in soils had its birth in the midst of the often acrimonious controversy that erupted in the 19th century concerning the mechanism of exchange processes. This controversy focused on the question of whether cation exchange was a kind of adsorption phenomenon, with the proponents declaring that such a mechanism precluded any theoretical description in terms of chemical equilibrium and the Law of Mass Action. This failure of the early investigators to understand that the principles of equilibrium among chemical species are macroscopic concepts not dependent on whatever molecular mechanism may prevail is easily appreciated, since chemical thermodynamics was not fully developed until the second decade of the present century. However, it is unfortunate that confusion between thermodynamics and molecular theory still persists here and there in the modern literature of soil chemistry, despite the advice given so clearly 29 years ago by Babcock et al. (1951):

> The first and second laws of thermodynamics were originally formulated independently of the atomic theory of matter. These laws, inductively discovered, have been found to describe all experimental situations investigated and, since they have never been contradicted, are generally accepted. The laws of thermodynamics concern themselves solely with directly observable quantities, and as a result, no rigorous inferences from them relevant to the atomic theory of matter are possible.

Regardless of how informative are data and models relating to the molecular mechanism of cation exchange, they are not directly germane to a purely thermodynamic description of the exchange phenomenon.

The earliest assertion in the published literature that cation exchange reactions can be described by some kind of equilibrium constant appears to be that of Ganssen (1913). He suggested (along with some sharp broadsides directed to the adherents of the adsorption hypothesis) that binary exchange processes on aluminosilicates could be characterized by a mass action coefficient:

$$K = c_1'(c_0 - c_1)/(s_0 - c_1')c_1 \qquad [1]$$

where c_1 is the millimolar concentration of cation "1" in aqueous solution, c_1' is the number of equivalents of cation "1" bound to unit mass of the exchange complex, c_0 is the total number of millimoles of cations per liter in aqueous solution, and s_0 is the cation exchange capacity. Ganssen (1913) (or Gans, as he is usually cited) does not seem to have based his equation on a clear picture of a cation exchange reaction, nor to have considered the thermodynamic implications of employing exchanger concentrations in equivalents per unit mass. It is, moreover, quite certain that Ganssen (1913) would not have anticipated the use of a thermodynamic exchange equilibrium constant expressed as an activity product, because the concept of thermodynamic activity was invented only that same year by Lewis (1913).

An important second step in the application of the principles of chemical equilibrium to cation exchange in soils was taken 15 years later by Kerr (1928). Although Kerr (1928) appears not to have known about Ganssen's mass action coefficient, he still cannot be accorded priority for his cation exchange coefficient, which is identical in form with Eq. [1]. Kerr's original contribution is his demonstration that the mass action coefficient in Eq. [1] reflects directly the fact that the soil exchange complex is a solution.

Suppose that a soil exchange complex which is saturated with magnesium is reacted with an aqueous solution of $CaCl_2$:

$$CaCl_2(aq) + MgX_2(s) = MgCl_2(aq) + CaX_2(s) \qquad [2]$$

where X represents one equivalent of the exchange complex. If one assumes that the two solid phases do not intermingle with one another to form a homogeneous mixture, Kerr (1928) reasoned, then their "active masses" will not vary with the composition of the exchange complex and the mass action coefficient which describes the reaction in Eq. [2] must reduce to:

$$K_1 = [MgCl_2]/[CaCl_2] \qquad [3]$$

where [ ] denotes an aqueous solution concentration. On the other hand, if the two solid phases mix homogeneously and, therefore, form a solid solution, their "active masses" are expected to vary proportionately with their total masses in the mixture and the mass action coefficient becomes:

$$K_2 = [MgCl_2]\{CaX_2\}/[CaCl_2]\{MgX_2\} \qquad [4]$$

Where { } denotes a concentration in the exchange complex. Table 1 shows some of Kerr's (1928) experimental data for the reaction in Eq. [2] as it occurred in Miami silt loam. It is quite evident that K_1, listed in the fifth column, depends strongly on the composition of the exchange complex, whereas K_2, in the sixth column, does not. The conclusion drawn by Kerr (1928), based on these and similar experimental results, was that soil exchangers behave as solid solutions and not as assemblies of separate solid phases.

The choice by Kerr (1928) of Mg-Ca exchange as the reaction on which to test his ideas was especially fortunate, since Mg-Ca exchangers often form nearly ideal solid solutions (a concept to be discussed below) and appear to have done so in every case that Kerr (1928) investigated. Had the soil exchange complexes Kerr (1928) studied been strongly non-ideal solid solutions, the values of K_2 he calculated could have been as de-

Table 1. Data obtained by H. W. Kerr (1928) for Ca-Mg exchange on Miami silt loam, demonstrating that the exchanger phase forms a solid solution.

[CaCl$_2$]	[MgCl$_2$]	{CaX$_2$}	{MgX$_2$}	$\dfrac{[MgCl_2]}{[CaCl_2]}$	$\dfrac{[MgCl_2]\{CaX_2\}}{[CaCl_2]\{MgX_2\}}$
—— mol dm^{-3} ——		—— mol kg^{-1} ——			
0.032	0.105	0.029	0.032	3.3	2.97
0.038	0.178	0.023	0.038	4.7	2.84
0.040	0.233	0.021	0.040	5.8	3.06
0.042	0.285	0.019	0.042	6.8	3.07
0.045	0.367	0.016	0.045	8.2	2.90
0.047	0.477	0.014	0.047	10.1	3.02
					Mean: 2.98 ± 0.07

pendent on exchanger composition as were those of K_1 and the sound, general conclusion which Kerr (1928) drew would not have been possible.

It is probable that Kerr (1928) meant "activity" when he wrote "active mass," although no reference to a work on chemical thermodynamics appears in his paper. Thus Kerr's development of Eq. [4] is the first treatment of cation exchange in soils which is truly thermodynamical in spirit, if not in provenance. Later in his paper, Kerr (1928) gives some consideration to the mass action coefficient for heterovalent cation exchange. This effort, however, produced ambiguous results because of Kerr's preoccupation with the peripheral question of whether the anionic part of the soil exchange complex was univalent or bivalent. Once again a confusion between macroscopic principles and molecular mechanisms had crept in.

Although Kerr (1928) might not have been expected to employ the concept of thermodynamic activity in his discussion of cation exchange reactions, the same cannot be said for Vanselow, who was a student of Lewis. Vanselow (1932) gave cation exchange its first modern representation in thermodynamic terms. He considered specific cases of the exchange reaction:

$$uBCl_v(aq) + vAX_u(s) = vACl_u(aq) + uBX_v(s) \qquad [5]$$

both theoretically and experimentally, where A^{u+} and B^{v+} are the two exchanging cations. The thermodynamic equilibrium constant which describes Eq. [5] is, according to Vanselow (1932):

$$K_{eq} = (ACl_u)^v(BX_v)^u/(BCl_v)^u(AX_u)^v \qquad [6]$$

where ( ) denotes a thermodynamic activity. [Actually, Vanselow (1932) writes $(B^{v+})^u$ and $(A^{u+})^v$ in place of the activities of the chloride salts, but his text makes it clear that he is thinking always of directly observable activities in his equations.] Vanselow (1932) pointed out that the experimental evaluation of K_{eq} in Eq. [6] depends on what expressions are chosen to relate the activities to concentration variables. In the case of the chloride salts, no special problem existed, because molal or molar concentration scales and the corresponding activity coefficients had been defined and measured for electrolytes in aqueous solution. In the case of the components of an exchanger phase, however, some ambiguity could arise because a standard thermodynamic concentration scale for exchanger phases had not yet been established.

One possibility, which Vanselow (1932) considered and rejected, was that the activity of an exchanger component was equal to the product of an activity coefficient and the number of moles (or equivalents) of the component per unit mass of exchanger. This choice is the same as that made by Kerr (1928) (along with the tacit assumption that all exchanger phase activity coefficients are equal). In Vanselow's (1932) opinion, such a choice was untenable because it treated exchanger components formally as if they were dissolved in an aqueous solution. The preference of Vanselow (1932) was to adopt some features of the thermodynamic theory of homogeneous mixtures, in which one always relates the mole fraction concentration scale to activity. Thus Vanselow (1932) wrote down the empirical exchange equilibrium constant:

$$K_V = (ACl_u)^v M_B{}^u / (BCl_v)^u M_A{}^v \qquad [7]$$

where

$$M_B = \frac{\{BX_v\}}{\{AX_u\} + \{BX_v\}} \qquad\qquad M_A = \frac{\{AX_u\}}{\{AX_u\} + \{BX_v\}}$$

are the mole fractions of $BX_v(s)$ and $AX_u(s)$, respectively, and $\{\ \}$ denotes specifically a concentration in moles per unit mass of exchanger.

The parameter K_V proposed by Vanselow (1932) was supposed to be the same numerically as the equilibrium constant, K_{eq} in Eq. [6]. Unfortunately, this is not generally true because of a very important fact which was overlooked by Vanselow (1932). The activity of a component of a homogeneous mixture is equal to its mole fraction only if the mixture is ideal (Guggenheim, 1967). If the mixture is not ideal, the activities of its components have to be related to their mole fractions by means of rational activity coefficients. Then Eq. [7] must be rewritten in the more general form:

$$K_{eq} = (ACl_u)^v f_B{}^u M_B{}^u / (BCl_v)^u f_A{}^v M_A{}^v \qquad [8]$$

where the exchanger activity coefficients are defined by the equations:

$$f_B \equiv (BX_v)/M_B \qquad\qquad f_A \equiv (AX_u)/M_A \qquad [9]$$

It is apparent, upon comparison of Eq. [6] with Eq. [8], that

$$K_V = K_{eq} f_A{}^v / f_B{}^u \qquad [10]$$

and, therefore, that K_V cannot properly be an equilibrium constant unless it so happens that $f_A = f_B = 1$, which is the necessary condition for a solid solution to be ideal. K_V in Eq. [7] is actually what should be termed a cation exchange selectivity coefficient. On the other hand, Eq. [6], due also to Vanselow (1932), is a correct and general equation for a binary exchange equilibrium constant.

EMPIRICALLY-BASED EXCHANGE CONSTANTS

The period during which Eq. [6] evolved was marked by a flurry of effort to develop a practicable and widely applicable mathematical relationship between the amount of a cation in an exchanger phase and the concentration of that cation at equilibrium in the aqueous solution phase. None of the many equations put forward enjoyed wide acceptance among soil chemists, as most of the expressions were based on specific (but sometimes unstated) assumptions concerning the interactions between exchangeable cations and the exchanger anion or the interactions between the exchangeable cations themselves. If an adsorption mechanism was assumed to be the cause of cation exchange, the mathematical connecting link between the exchanger phase and the aqueous solution with which it made contact was a chosen adsorption isotherm equation. If mass action concepts were favored instead, some kind of ansatz for the exchanger phase activities was invoked. Regardless of the underlying motivation, the result of any empirically-based exchange model can be given an interpretation in chemical thermodynamics with the help of Eq. [8].

For example, suppose that the exchange reaction in Eq. [5] is under consideration for the case $u = v$ (homovalent exchange) and it is assumed that the concentration of BX_v in the exchanger phase and that of BCl_v in aqueous solution are related according to the adsorption isotherm equation:

$$M_B = \alpha[BCl_v]^\beta \qquad [11]$$

where α and β are empirical parameters. Then the mass balance relations

$$M_B + M_A = 1 \qquad c_T = [ACl_v] + [BCl_v] \qquad [12]$$

where c_T is the total molar cation concentration, can be applied to derive an equation for the Vanselow selectivity coefficient:

$$
\begin{aligned}
K_V &= (ACl_v)^v M_B{}^v/(BCl_v)^v (1 - M_B)^v \\
&= [ACl_v]^v M_B{}^v/[BCl_v]^v (1 - M_B)^v \\
&= (c_T - [BCl_v])^v M_B{}^v/[BCl_v]^v (1 - M_B)^v \\
&= [M_B(c_T - \alpha^{-1/\beta} M_B{}^{1/\beta})/\alpha^{-1/\beta} M_B{}^{1/\beta} (1 - M_B)]^v \\
&= [M_B(R - M_B{}^{1/\beta})/M_B{}^{1/\beta} (1 - M_B)]^v \qquad [13]
\end{aligned}
$$

where $R = c_T \alpha^{1/\beta}$. In deriving Eq. [13], the assumption has been made that the molar activity coefficients of the two chloride salts are equal. Equation [11] is an adsorption isotherm equation first suggested by van Bemmelen (1888), although it is often attributed to Freundlich (1909).

The particular K_V that follows from van Bemmelen's (1888) equation will vary with the composition of the exchange complex unless the empirical parameters R and β both happen to have unit value. Therefore, except for special cases, Eq. [11] describes a non-ideal exchanger in equilibrium with an aqueous solution. This fact may be summarized mathematically by combining Eq. [10] (with u = v) and Eq. [13] to produce the expression:

$$K_{eq}^{1/v}\frac{f_A}{f_B} = \frac{M_B(R - M_B^{1/\beta})}{M_B^{1/\beta}(1 - M_B)} \qquad [14]$$

It is evident that f_A and f_B do not equal to 1.0 and that the dependence on M_B of the right-hand side of Eq. [14] is proportional to that of the activity coefficient ratio, f_A/f_B, since K_{eq} is a constant. The thermodynamic implication of van Bemmelen's (1888) equation (Eq. [11]) is that it amounts to a specification of the ratio of exchanger phase activity coefficients. This same general conclusion may be applied to any other empirically-derived adsorption isotherm equation.

Table 2, which is patterned after a study by Högfeldt (1955), shows five examples of K_V calculated according to some empirically-based prescription for the composition of an exchanger phase undergoing uni-univalent exchange (u = v = 1 in Eq. [5]). The first three expressions for K_V result from applications of Eq. [11]. For example, the Wiegner-Jenny expression is the varient of the van Bemmelen-Freundlich equation which is obtained by writing $\beta = q/(1+q)$ (Wiegner and Jenny, 1927). The Rothmund and Kornfeld (1918) equation is derived by applying Eq. [11] separately to the exchanger components BX and AX, under the assumption that β is the same for both cations while α is not. Then only the first of the mass balance Eq. [12] need be employed to calculate K_V. On the

Table 2. Five empirically-based theories that produce a Vanselow selectivity coefficient describing the reaction: BCl(aq) + AX(s) = ACl(aq) + BX(s).

Author	Expression for K_V
van Bemmelen/Freundlich	$K_V = \dfrac{M_B(R - M_B^{1/\beta})}{M_B^{1/\beta}(1 - M_B)}$
Wiegner and Jenny	$K_V = \dfrac{(R - M_B^{1+\frac{1}{q}})}{M_B^{1/q}(1 - M_B)}$
Rothmund and Kornfeld	$K_V = k^{1/\beta}\dfrac{M_B(1 - M_B)^{1/\beta}}{M_B^{1/\beta}(1 - M_B)}$
Vageler and Woltersdorf	$K_V = \dfrac{[1 - (1 + Q(\sigma - s_o))M_B + s_o M_B]\dagger}{Q[(\sigma - s_o) + M_B](1 - M_B)}$
Kielland	$K_V = K_{eq}\exp[C(M_B^2 - (1 - M_B)^2)]$

† $Q\sigma = c_{B\frac{1}{2}}/c_T$ (see text). s_o = cation exchange capacity.

other hand, the Vageler and Woltersdorf (1930) equation for K_V develops from the empirical relationship:

$$M_B = [BCl]_o/(c_{B^{1/2}} + [BCl]_o) \qquad [15]$$

where $[BCl]_o$ is the concentration of BCl added at the beginning of the exchange reaction and c_B is the value of $[BCl]_o$ which makes $M_B = 0.5$. Finally, the Kielland (1935) equation is based on the assumption that $f_B = \exp(+CM_A^2)$ and $f_A = \exp(+CM_B^2)$, where C is an empirical constant.

None of the expressions for K_V in Table 2 are of universal applicability (Högfeldt, 1955), but each serves to illustrate how empirical postulates concerning the distribution of cations between an exchanger phase and an aqueous solution can be interpreted in a unified fashion through chemical thermodynamics. In a similar vein, the many ad hoc equations which have been proposed to express cation exchange equilibrium constants can be incorporated into a general thermodynamic framework by noting that they, too, are equivalent to specific assumptions about the composition dependence of exchanger phase activity coefficients. Some examples of empirically-based expressions for cation exchange equilibrium constants, proposed to describe the uni-bivalent exchange reaction:

$$BCl_2(aq) + 2 AX(s) = 2 ACl(aq) + BX_2(s) \qquad [16]$$

are given in Table 3. In each case, the mathematical form of K_{eq} can be derived from a combination of the expression:

$$K_{eq} = (ACl)^2 f_B M_B/(BCl_2)f_A^2 M_A^2 \qquad [17]$$

with a specific assumption about the composition dependence of the activity coefficients of the components of the exchanger phase. The well-

Table 3. Some examples of empirical exchange equilibrium constants which have been proposed to describe the reaction: $BCl_2(aq) + 2 AX(s) = 2 ACl(aq) + BX_2(s)$.

Author	Equilibrium constant	Activity relations assumed	Activity coefficients
Vanselow	$K_V =$ $(ACl)^2 M_B/(BCl_2)M_A^2$	$(AX) = M_A$ $(BX_2) = M_B$	$f_A = 1$ $f_B = 1$
Gaines and Thomas	$K_{GT} =$ $(ACl)^2 E_B/(BCl_2)E_A^2$	$(AX) = E_A$[†] $(BX_2) = E_B$	$f_A = (2-M_A)^{-1}$ $f_B = 2(1-M_B)^{-1}$
Rothmund and Kornfeld	$K_{RK} =$ $(ACl)^2 E_B^{2/\beta}/(BCl_2)E_A^{2/\beta}$	$(AX) = E_A^{1/\beta}$ $(BX_2) = E_B^{2/\beta}$	$f_A = M_A^{1/\beta-1}(2-M_A)^{-1/\beta}$ $f_B = 4^{1/\beta}M_B^{2/\beta-1}(1+M_B)^{-2/\beta}$
Gapon	$K_G{}^2 =$ $(ACl)^2 E_B^2/(BCl_2)E_A^2$	$(AX) = E_A$ $(BX_2) = E_B^2$	$f_A = (2-M_A)^{-1}$ $f_B = 4M_B(1+M_B)^{-2}$

† $E_B = 2 M_B/(M_A + 2 M_B)$ and $E_A = M_A/(M_A + 2 M_B)$ are equivalent fractions of BX_2 and AX, respectively.

known Gapon (1933) equation is the special case of the earlier Rothmund and Kornfield (1919) equation which results for $\beta = 1$. The Gaines and Thomas (1953) equation is sometimes called the Kielland equation (Barrer and Klinowski, 1974), but this designation does not seem to be justified, since Kielland (1935) did not discuss heterovalent exchange nor even mention equivalent fractions.

THE GENERAL THERMODYNAMIC DESCRIPTION OF CATION EXCHANGE EQUILIBRIA

Eighteen years after Vanselow (1932) expressed the exchange equilibrium constant as a ratio of activities, Argersinger et al. (1950) and Högfeldt (Ekedahl et al., 1950; Högfeldt et al., 1950a) independently derived a set of general thermodynamic equations which related exchanger phase activity coefficients to exchanger phase composition. These equations reduced the computation of exchanger phase activities and, therefore, the calculation of K_{eq}, to a problem of quadrature that required no more than the measurement of the Vanselow selectivity coefficient as a function of exchanger phase composition. Chemical thermodynamics thus was shown to provide a complete, self-consistent description of cation exchange based solely in the macroscopic concepts of the mole fraction, the molality or molarity, and the activity coefficient.

Argersinger et al. (1950) presented a derivation of the equations for the activity coefficients of $AX_u(s)$ and $BX_v(s)$ in Eq. [5] which was a model of economy. These investigators saw clearly that the essential content of Eq. [10] is the fact that any variation in K_V with respect to exchanger phase composition must be the result of variation in the exchanger phase activity coefficients:

$$v \, d\ln f_A - u \, d\ln f_B = d\ln K_V \qquad [17]$$

where natural logarithms of both sides of Eq. [10] have been taken before calculating the differentials. Secondly, Argersinger et al. (1950) noted that any variation in the activity of $AX_u(s)$ must be compensated for by a variation in the activity of $BX_v(s)$ in such a way that mass in the exchanger phase is conserved. This requirement, an application of the Gibbs-Duhem equation (Guggenheim, 1967), produces the equation:

$$M_A \, d\ln f_A + M_B \, d\ln f_B = 0 \qquad [18]$$

Equations [17] and [18] are a set of two algebraic equations in the logarithmic differentials of the two unknown activity coefficients, f_A and f_B. The solutions of these equations are:

$$v \, d\ln f_A = E_B \, d\ln K_V \qquad u \, d\ln f_B = -(1 - E_B) d\ln K_V \qquad [19]$$

where $E_B = vM_B/(uM_A + vM_B)$ is the equivalent fraction of $BX_v(s)$ in the exchanger phase. The integration of Eq. [19] provides the values of $\ln f_A$ and $\ln f_B$ at any value of E_B:

$$v \ln f_A = \int_0^{E_B} E_B' \, d\ln K_V = \ln K_V - \int_0^{E_B} \ln K \, dE_B' \qquad [20a]$$

$$-u \ln f_B = -\int_{E_B}^1 (1 - E_B') d \ln K_V = (1 - E_B) \ln K_V - \int_{E_B}^1 \ln K_V \, dE_B' \qquad [20b]$$

In Eq. [20], integration by parts has been carried out and the assumptions that $f_A = 1$ when $E_B = 0$ and $f_B = 1$ when $E_B = 1$ have been made. Finally, since

$$\ln K_{eq} = \ln K_V - v \ln f_A + u \ln f_B \qquad [21]$$

according to Eq. [10], Eq. [20] lead to the expression:

$$\ln K_{eq} = \int_0^1 \ln K_V \, dE_B \qquad [22]$$

Ekedahl et al. (1950) and Högfeldt et al. (1950a) also derived Eq. [19], first for the case of homovalent exchange, then for heterovalent exchange. Högfeldt et al. (1950a) went futher than Argersinger et al. (1950), however, by demonstrating that thermodynamics allows an alternate formulation of the cation exchange reaction in Eq. [5]. The cations may be designated formally to react with the exchanger anion in equivalents instead of in moles, with no violation of the macroscopic laws of mass and charge conservation:

$$uBCl_v(aq) + uvA_{\underset{u}{1}}X(s) = vACl_u(aq) + uvB_{\underset{v}{1}}X(s) \qquad [23]$$

In this case, Eq. [6] becomes the expression:

$$K_{eq} = (ACl_u)^v (B_{\underset{v}{1}}X)^{uv}/(BCl_v)^u (A_{\underset{u}{1}}X)^{uv} \qquad [24]$$

and Eq. [7] may be replaced by the equation:

$$K_H = (ACl_u)^v E_B^{uv}/(BCl_v)^u E_A^{uv} \qquad [25]$$

where E again denotes an equivalent fraction. The expression corresponding to Eq. [10] is then:

$$K_H = K_{eq} h_A^{uv}/h_B^{uv} \qquad [26]$$

where

$$h_B \equiv (B_{\underset{v}{1}}X)/E_B \qquad h_A \equiv (A_{\underset{u}{1}}X)/E_A \qquad [27]$$

Högfeldt et al. (1950a) employed Eq. [26] and the Gibbs-Duhem equation to derive equations analogous to Eq. [19]. If these equations are integrated, the analogs of Eq. [20] and [22] result.

All of the manipulations carried out by Högfeldt et al. (1950a) are correct formally, in that K_H is a selectivity coefficient [a generalization of the Gapon (1933) exchange equilibrium constant] which may be integrated in order to calculate the exchange equilibrium constant, K_{eq}. However, it was overlooked by Högfeldt et al. (1950a) that, except in the case of homovalent exchange, the quantities h_B and h_A, defined in Eq. [27], are not exchanger phase activity coefficients. An activity coefficient is, by definition, always the ratio of the actual activity to the value of the activity under those limiting conditions when Raoult's Law applies (Sposito and Mattigod, 1979; Whiffen, 1979). Therefore, in solid solutions, such as exchanger phases, an activity coefficient always is the ratio of an actual activity to a mole fraction, as indicated in Eq. [9]. The ratio of an activity to an equivalent fraction is only a formal parameter which cannot be interpreted directly in terms of a thermodynamic reference state, as can f_A and f_B. The reason that the equivalent fraction concentration scale is not used to define an activity coefficient in an exchanger phase is precisely the same as the reason a normality concentration scale is never used to define an activity coefficient in an aqueous solution (Sposito, 1977). Equivalents are purely formal quantities which are not associated with actual chemical species, except in the special case of univalent ions. The failure to appreciate this basic chemical fact has led to much confusion in the experimental application of thermodynamics to cation exchange reactions, beginning even with the work of Högfeldt et al. (1950b). For example, the parameters h_B and h_A in Eq. [27] can become infinite or vanish in an exchanger whose chemical behavior is actually quite normal (Högfeldt, 1953; Sposito, 1977). The cause of this unusual result is purely mathematical, not chemical.

The thermodynamic description of cation exchange equilibria was given further consideration by Gaines and Thomas (1953), who made two important contributions. First, they brought up the fact that, in some exchangers, significant amounts of water are adsorbed and the activity of that water may change as the composition of the exchanger phase is varied. This possibility can be taken into account by including a term in the water activity when the Gibbs-Duhem equation is written down for the exchanger phase in the process of deriving Eq. [18]. The term to be added to Eq. [18] is $n_w \, dln \, a_w$, where n_w is the number of moles of water in the exchanger per total moles of exchangeable cations and a_w is the activity of water in the exchanger phase. The need for such a term actually was seen independently by Högfeldt (1953), Davidson and Argersinger (1953), and Gaines and Thomas (1953). However, as has been discussed in detail by Holm (1956), the treatment of the water activity term by Gaines and Thomas (1953) involved the least restrictive assumptions concerning the nature of water adsorption by an exchanger. Recent work (Laudelout and Thomas, 1965; Bolt and Winkelmolen, 1968; Laudelout et al., 1971; Barrer and Klinowski, 1974) has indicated that changes in the water activity as the exchanger composition is varied have no significant effect on the computation of K_{eq}, although they can be important in the calculation of exchanger phase activity coefficients, at least in the case of zeolitic exchangers (Barrer and Klinowski, 1974).

The second, and more significant, contribution made by Gaines and Thomas (1953) was to reiterate in concrete terms a fact discussed by Babcock et al. (1951) concerning the reference state in which exchanger phase activity coefficients are defined to have unit value. Both Argersinger et al. (1950) and Högfeldt et al. (1950a) chose the exchanger phase activity coefficient to be equal to one when the mole fraction of the component it refers to was equal to one. However, as indicated by Babcock et al. (1951) and discussed later by Holm (1956), this specification of the reference state is not complete because the properties of an exchanger phase will depend, in general, on the ionic strength of the aqueous solution as well as on the nature of the exchange complex itself. If a set of otherwise identical cation exchange experiments is carried out at different ionic strengths, it can be expected that the exchanger phase activity coefficients for each set, computed with the help of Eq. [20], will be different even at the same value of E_B. Gaines and Thomas (1953) provided a resolution of this problem by stating that the reference state of a component of an exchanger phase is the homoionic exchanger consisting of that component in equilibrium with an infinitely dilute, aqueous solution containing the exchangeable cation.

In practice, it appears that exchanger phase activity coefficients in soils and clays show only a very slight dependence on the ionic strength of the bathing electrolyte solution (Jensen, 1973; Jensen and Babcock, 1973). This fact may serve as a justification for the almost universal neglect of the Gaines-Thomas reference state when experimental values of exchanger phase activity coefficients actually are reported. But it is worthwhile to remember that the left-hand sides of Eq. [20] and [22] are in reality functions of ionic strength unless f_A and f_B are corrected to infinite dilution conditions.

Gaines and Thomas (1953) chose, unfortunately, to write the exchange equilibrium constant of the reaction in Eq. [5] in the form:

$$K_{eq} = (ACl_u)^v g_B^u E_B^u / (BCl_v)^u g_A^v E_B^v \qquad [28]$$

instead of as in Eq. [8]. The parameters g_A and g_B are defined by the expressions:

$$g_B \equiv (BX_v)/E_B \qquad g_A \equiv (AX_u)/E_A \qquad [29]$$

Accordingly, the calculation of K_{eq} presented by Gaines and Thomas (1953) employed the selectivity coefficient

$$K_{GT} = (ACl_u)^v E_B^u / (BCl_v)^u E_A^v \qquad [30]$$

and not K_V in Eq. [7]. It follows that the Gaines-Thomas calculation of K_{eq} is only a formal procedure and that, except for homovalent exchanges, g_B and g_A in Eq. [29], like h_B and h_A in Eq. [27], are only formal parameters without a strict thermodynamic meaning in themselves. This is a most unfortunate circumstance, since measured values of g_B and g_A have been published for many heterovalent cation exchange reactions in soils and clays and often have been interpreted as if they were exchanger phase activity coefficients (Sposito and Mattigod, 1979).

CONCLUSIONS

The historical development of the thermodynamic theory of cation exchange in soils can be viewed as a search along a somewhat tortuous path for a general, self-consistent, macroscopic description of exchanger phase activity coefficients. It is into these parameters and the exchange equilibrium constant that chemical thermodynamics lumps all of the complexity of the mutual interactions that go on among the components of an exchanger, and it is these parameters alone which are the proper object of any thermodynamic or molecular theory of exchange reactions.

Perhaps it is useful to add a word at this point concerning the molecular theory of cation exchange, whose object is to derive the thermodynamic properties of cation exchange reactions from the principles of statistical mechanics. The molecular theory seeks to explain how the underlying atomic configurations and interactions in a soil exchange complex produce the observed values of exchange equilibrium constants and the observed composition dependence of the activity coefficients of exchanger components. Thermodynamic descriptions of soil systems can do no more than introduce quantities like equilibrium constants and activity coefficients solely as empirical parameters. The molecular behavior in a soil which results in the actual, measured values of these quantities is properly the object of statistical mechanics to describe. The practice of molecular theory involves the construction of simplified models (e.g., the diffuse double layer model) which hopefully will make the ab initio calculation of thermodynamic properties a tractable mathematical problem without a total sacrifice of chemical reality.

ACKNOWLEDGMENTS

Professor A. L. Page suggested that I undertake this review and most kindly read the manuscript in draft. I am much indebted to Ms. Nancy Ball for her careful reading of this paper and for numerous suggestions concerning its content and style. My deepest gratitude for constant encouragement, unfailing support, and perceptive guidance goes to Professor K. L. Babcock, to whom this paper is dedicated with much affection.

LITERATURE CITED

1. Argersinger, W. J., A. W. Davidson, and O. D. Bonner. 1950. Thermodynamics and ion exchange phenomena. Trans. Kansas Acad. Sci. 53:404–410.

2. Babcock, K. L. 1963. Theory of the chemical properties of soil colloidal systems at equilibrium. Hilgardia 34:417–542.

3. ————, L. E. Davis, and R. Overstreet. 1951. Ionic activities in ion-exchange systems. Soil Sci. 72:253–260.

4. Barrer, R. M., and J. Klinowski. 1974. Ion-exchange selectivity and electrolyte concentration. J. Chem. Soc. Faraday Trans. I 70:2,080–2,091.

5. van Bemmelen, J. M. 1888. Die Absorptions Verbindungen und das Absorptions Vermögen der Ackererde. Landwirtschaft Vers. Stat. 35:69–136.

6. Bolt, G. H. 1967. Cation-exchange equations used in soil science—a review. Neth. J. Agric. Sci. 15:81–103.

7. ————, and C. J. G. Winkelmolen. 1968. Calcualtion of the standard free energy of cation exchange in clay systems. Israel J. Chem. 6:175–187.

8. Davidson, A. W., and W. J. Argersinger. 1953. Equilibrium constants of cation exchange processes. Ann. New York Acad. Sci. 57:105–115.

9. Ekedahl, E., E. Högfeldt, and L. G. Sillén. 1950. Activities of the components in ion exchangers. Acta Chem. Scan. 4:556–558.

10. Forster, E. S. 1927. The works of Aristotle. Vol. VII. Problemata. Oxford Univ. Press, London.

11. Freundlich, H. 1909. Kappillarchemie. Leipzig.

12. Gaines, G. L., and H. C. Thomas. 1953. Adsorption studies on clay minerals. II. A formulation of the thermodynamics of exchange adsorption. J. Chem. Phys. 21:714–718.

13. Ganssen (Gans), R. 1913. Über die chemische oder physikalische Natur der kolloidalen wasserhaltigen Tonerdesilicate. Zentralblatt fur Mineralogie, Geologie Paläontologie 699–712, 728–741.

14. Gapon, Y. N. 1933. On the theory of exchange adsorption in soils. J. Gen. Chem. USSR 3:144–160.

15. Guggenheim, E. A. 1967. Thermodynamics. North-Holland, Amsterdam.

16. Helfferich, F. 1962. Ion exchange. McGraw-Hill, New York.

17. Högfeldt, E. 1953. On ion exchange equilibria. II. Activities of the components in ion exchangers. Arkiv Kemi 5:147–171.

18. ————. 1955. On ion exchange. III. An investigation of some empirical equations. Acta Chem. Scand. 9:151–165.

19. ————, E. Ekedahl, and L. G. Sillén. 1950a. Activities of the components in ion exchangers with multivalent ions. Acta Chem. Scand. 4:828–829.

20. ————, ————, and ————. 1950b. Activities of the barium and hydrogen forms of Dowex 50. Acta Chem. Scand. 4:829–830.

21. Holm, L. W. 1956. On the thermodynamics of ion exchange equilibria. I. The thermodynamical equilibrium equation in relation to reference states and components. Arkiv Kemi 10:151–166.

22. Jenny, H. 1932. Studies on the mechanism of ionic exchange in colloidal aluminum silicates. J. Phys. Chem. 36:2,217–2,258.

23. Jensen, H. E. 1973. Potassium-calcium exchange equilibria on a montmorillonite and a kaolinite clay. I. A test on the Argersinger thermodynamic approach. Agrochimica 17:181–190.

24. ————, and K. L. Babcock. 1973. Cation-exchange equilibria on a Yolo loam. Hilgardia 41:475–488.

25. Kelley, W. P. 1948. Cation exchange in soils. Reinhold Publ. Corp., New York.

26. Kerr, H. W. 1928. The nature of base exchange and soil acidity. J. Am. Soc. Agron. 20:309–335.

27. Kielland, J. 1935. Thermodynamics of base-exchange equilibria of some different kinds of clays. J. Soc. Chem. Ind. 54:232T–234T.

28. Laudelout, H., and H. C. Thomas. 1965. The effect of water activity on ion-exchange selectivity. J. Phys. Chem. 69:339–341.

29. ————, R. van Bladel, and J. Robeyns. 1971. The effect of water activity on ion exchange selectivity in clays. Soil Sci. 111:211–213.

30. Lewis, G. N. 1913. The free energy of chemical substances. Introduction. J. Am. Chem. Soc. 35:17–30.

31. Reddy, M. M. 1977. Ion-exchange materials in natural water systems. Ion Exch. Solvent Extr. 7:165–219.

32. Reichenberg, D. 1966. Ion-exchange selectivity. Ion Exch. Solvent Extr. 1:227–276.

33. Rothmund, V., and G. Kornfield. 1918. Die Basenaustausch im Permutit. I. Z. Anorg. und Allg. Chem. 103:129–163.

34. ————, and ————. 1919. Die Basenaustausch im Permutit. II. Z. Anorg. und Allg. Chem. 108:215–225.

35. Sposito, G. 1977. The Gapon and the Vanselow selectivity coefficients. Soil Sci. Soc. Am. J. 41:1205–1206.

36. ————, and S. V. Mattigod. 1979. Ideal behavior in Na-trace metal cation exchange on Camp Berteau montmorillonite. Clays Clay Min. 27:125–128.

37. Thomas, G. W. 1977. Historical developments in soil chemistry: Ion exchange. Soil Sci. Soc. Am. J. 41:230–238.

38. Vageler, P., and J. Woltersdorf. 1930. Beiträge zur Frage des Basenaustausches und der Aziditaten. Z. Pflanzenernahr. Düngung. A15:329–342.

39. Vanselow, A. P. 1932. Equilibria of the base-exchange reactions of bentonites, permutites, soil colloids, and zeolites. Soil Sci. 33:95–113.

40. Whiffen, D. H. 1979. Manual of symbols and terminology for physiochemical quantities and units. Pure Appl. Chem. 51:1–41.

41. Wiegner, G. 1912. Zum Basenaustausch in der Ackererde. Journ. Landw. 60:111–150, 197–222.

42. ————, and H. Jenny. 1927. Über Basenaustausch an Permutiten. Kolloid-Zietschrift 42:268–272.

QUESTIONS AND ANSWERS

Q. Are we at the point where we should be able to standardize the exchange equilibrium constant so that we could use it? And how would we do this?

A. Yes. If one follows the Argersinger procedure (probably it is okay to neglect the water activity contribution), then, in fact, one is calculating things of thermodynamic significance. One can also use molecular theories, such as the diffuse double layer theory or some of the site-binding models, to try to calculate the activity coefficients, ab initio, if one wishes. On the other hand, one does not need to do that because it is possible to get those activity coefficients solely from the experimentally measured information. These coefficients could be tabulated and trends could be looked for among different soils, and so forth. As a matter of fact, some analysis of this type is already possible with the data for individual clay minerals.

Q. This will give us a characterization of soils?

A. Yes. Actually, a parallel thing has already occurred in solid solution geochemistry. All these thermodynamic equations apply there too, since, as I mentioned, this analysis is independent of the exchange mechanism.

Q. You propose a thermodynamic treatment for the individual ions in the reaction. In view of the current understanding of the soil solution chemistry, how would you define the different ion species? Unless you know the effective ion-size parameters, you cannot calculate the activity coefficients. How does this affect the use of thermodynamic activities for the individual ions?

A. Not at all. The aqueous solution phase can be modelled fairly effectively by the use of the Davies equation for the single-ion activity coefficients in a certain range of ionic strength. This does not require the Debye-Hückel equation. I personally would advise against the use of the Debye-Hückel equation in any soil solution, although it is widely used. At any rate, the aqueous solution phase in soil can be modelled reasonably well. It is, afterall, a multi-component chemical system which can be handled on a computer. There is a number of multi-component chemical equilibrium computer programs available that can handle soil solution equilibria, certainly for most of the interesting ion exchange reactions. Typically we find that soil solutions contain around 350 to 400 soluble complexes. It takes about a minute or so of computer time to calculate their equilibrium concentrations. So the problem to which you are referring is primarily a numerical one, at least for the order of magnitude of accuracy that one wants for most problems of interest in soil chemistry. Nonetheless, the formulation of the thermodynamic theory of exchange does not require any particular information about the aqueous solution phase (insofar as to what you are referring goes) for its formulation. Now, your question is properly related to the application of the equations. Of course, the application of thermodynamics is always a difficult problem. The principles are simple. The applications are always difficult because one has to think clearly about what one is doing and has to have correct input information. But it certainly can be done. We do it, other people do it. Basically, soils are multicomponent chemical systems; the principles of chemistry apply. It just takes a little effort, that's all.

Q. Do we have all the parameters?

A. Well, we believe that we have enough parameters to begin to do work on the solutions in which we are interested. Others can do what they will, I guess. But the point is that the alternatives are to do nothing or to do something. I'll choose to do something! Yes, it is certainly going to be difficult. The soil solution is as challenging a chemical system as the ocean. It is probably going to take a long time to get everything pinned down, but I think we have to start.

Q. You mean that in a system such as ion-H_2O we can model effectively?

A. We don't know all the species, especially hydrolytic species. What one has to do, I think, is just begin to work on these systems, make calculations, estimates, measurements, etc., then compare them with information that we can get from other sources, and keep going. There's no real alternative. Basically, we are dealing with a problem of "chemical ecology." It is as complicated as the problem of animal ecology or plant ecology that biologists have to deal with. One advantage that we do have is that all the laws are known. All we have to do is find the species!

Q. You allude to the fact that geochemists are already calculating activity coefficients in pure systems. But I wish you would expand a little bit on the methods you suggest for us to calculate these activity coefficients.

A. This kind of calculation has been done a number of times for clay minerals. The work of Gast, for example, comes to mind immediately. It is a good example of a proper calculation for the monovalent exchanges with which he was dealing. It has been done correctly in some cases for heterovalent exchanges. The work of Jensen and Babcock, for example, comes to mind.

Q. They're all empirical; is that right?

A. No. One uses the thermodynamic expressions which I gave in the talk: one measures the Vanselow selectivity coefficient and integrates its natural logarithm with respect to the equivalent fraction. Then, out come the activity coefficients. With those, one has a set of empirical data. Now, the next step is to begin to take a look at lots of these data for a number of pure clay mineral systems and to see if trends can be established. There are also model expressions, used for years by geochemists to fit activity coefficients data in their solid-solution systems, which possibly could apply in the systems with which soil scientists deal. For that matter, this exercise could be done for whole soils because thermodynamics doesn't care. It simply is characterizing a system as a whole on a macroscopic level. One could then begin to look for trends in the activity coefficients, just like Davies did with aqueous solution systems to come up with the Davies equation. Maybe there is an analog of the Davies equation for exchangers in soil systems. We don't know yet.

CHAPTER 3

Principles of Ion Diffusion in Clays[1]

PHILIP F. LOW[2]

ABSTRACT

The basic equations that describe diffusion in a charged porous medium are presented. Also the factors that affect the coefficients and variables in these equations are identified and discussed. Experimental evidence is provided to support the concepts that are developed and the conclusions that are drawn.

The author has been interested in the principles of ion movement in clays since the beginning of his professional career when he published one of the first papers written on the subject (8). Subsequently, he has published many additional papers relating to it (1, 4, 5, 9, 12, 13, 14, 15, 17, 18, 20, 21, 22). These papers, which depend on earlier, theoretical treatments of conductance and diffusion (6, 10, 24), provide the basis for the present discussion. The information that they contain will be summarized and synthesized in order to develop definite concepts. The author does not intend to write a general review and so the work of others will be cited only if it helps to elucidate these concepts or contains relevant data.

Let us consider any ionic species, i, which is moving through a clay-water system in which no convective flow of water is occurring. If we assume that the driving force acting on the ions is the gradient of $\overline{G}_i$, their partial molar free energy, and that their average velocity, v_i, is proportional to this driving force, then

$$v_i = -\lambda_i \, (d\overline{G}_i/dy') \qquad [1]$$

[1] Journal paper no. 7833, Purdue Univ. Agr. Exp. Sta., West Lafayette, IN 47907. Contribution from the Agronomy Dep.

[2] Professor of soil chemistry.

in which y' is distance in a direction parallel to the surfaces of the clay particles and λ_i is a mobility factor that equals the average velocity attained when $(d\overline{G}_i/dy')$ is unity. The flux, j_i, of the species through an elementary plane having an infinitesimal width, dx', in a direction perpendicular to the particle surfaces and a height of 1 cm is

$$j_i = c_i v_i dx' \qquad [2]$$

where c_i is the ionic concentration. Combination of Eqs. [1] and [2] yields

$$j_i = -c_i \lambda_i \, dx' \, (d\overline{G}_i/dy'). \qquad [3]$$

Now, both c_i and λ_i vary with distance from the particle surfaces, i.e., in the x' direction. However, as shown earlier (20), Eq. [3] can be integrated over the distance, d, between any pair of adjacent particles and the result of the integration can then be summed for all such pairs in the system to obtain

$$J_i = -A \beta \overline{c}_i \overline{\lambda}_i (d\overline{G}_i/dy) \qquad [4]$$

in which J_i is the total flux of the species in the y direction (which is the direction of net movement), A is the total cross-sectional area perpendicular to this direction, β is a geometry factor that corrects A and dy for porosity and tortuosity, respectively, $\overline{c}_i$ is the average ionic concentration in the space between the particles and $\overline{\lambda}_i$, the average value of λ_i in this space, is defined by

$$\overline{\lambda}_i = \int_0^d c_i \lambda_i \, dx'/10^3 \, N \qquad [5]$$

where N is the number of moles of the species between the particles per cm^2 of opposing surface area. Also, as shown earlier (20),

$$\overline{G}_i = \overline{G}_i^\circ + RT \ln \overline{\gamma}_i \overline{c}_i \qquad [6]$$

where $\overline{G}_i^\circ$ is the partial molar free energy of the species in its standard state, R is the molar gas constant, T is the absolute temperature and $\overline{\gamma}$ is the space average activity coefficient. Differentiation of Eq. [6] with respect to y and substitution of the result into Eq. [4] gives

$$J_i = -A \beta \overline{\lambda}_i RT (d\overline{c}_i/dy) \qquad [7]$$

if $\overline{\gamma}$ does not change in the y direction. When infinitesimal concentrations of isotopes are used to study self-diffusion, there is essentially no change in composition along the diffusion path and so $\overline{\gamma}$ does not vary with y. Hence, Eq. [7] is applicable to self-diffusion. If this equation is compared to Fick's law for a single ionic species, viz.,

$$J_i = -A D_i (d\overline{c}_i/dy) \qquad [8]$$

where D_i is the diffusion coefficient, we can see that

$$D_i = \beta \bar{\lambda}_i RT. \tag{9}$$

If the concentration gradient of the ionic species is zero but it is subjected to an applied electrical potential gradient, $d\psi/dy$, then $d\overline{G}_i/dy'$ in Eq. [1] becomes $z_i F(d\psi/dy')$ and λ_i is replaced by $u_i/z_i F$, where u_i is the ionic mobility, z_i is the ionic valence and F is the faraday. The resulting equation can be combined with Eq. [2] and integrated as before to obtain

$$J_i = -A\beta \,\bar{c}_i\bar{u}_i(d\psi/dy) \tag{10}$$

in which $\bar{u}_i$ is the space average ionic mobility. Attention is called to the fact that Eq. [10] becomes invalid if the applied electrical potential gradient causes a significant electro-osmotic flow of water because this gradient is measured with respect to a fixed frame of reference, whereas, the motion of the ions must be measured with respect to the surrounding water. Observe that all the variables in the equation are measurable except β and the $\bar{u}_i$. We will define a quantity, U_i, by the relation

$$U_i = \beta \,\bar{u}_i \tag{11}$$

and call it the apparent mobility.

In any clay-water system, electroneutrality must be maintained and so no ionic species can move independently of the other ionic species that are present. For example, consider the diffusion of a single symmetrical electrolyte. Electroneutrality requires that the total charge transferred by the cations and anions, designated by the subscripts 1 and 2, respectively, be equal. Consequently,

$$z_1 J_1 = -z_2 J_2. \tag{12}$$

By combining Eq. [4], [6], [12] and the following thermodynamic expression

$$\overline{G}_{12} = \overline{G}_1 + \overline{G}_2 \tag{13}$$

where the double subscript denotes the electrolyte, we find (20) that

$$J_{12} = -A\beta \, RT \, \frac{\bar{\lambda}_1\bar{\lambda}_2(\bar{c}_1 + \bar{c}_2)}{(\bar{c}_1\bar{\lambda}_1 + \bar{c}_2\bar{\lambda}_2)} \, \frac{d\bar{c}_{12}}{dy}. \tag{14}$$

Comparison of this equation with Fick's law for an electrolyte shows that

$$D_{12} = \beta \, RT \, \frac{\bar{\lambda}_1\bar{\lambda}_2(\bar{c}_1 + \bar{c}_2)}{(\bar{c}_1\bar{\lambda}_1 + \bar{c}_2\bar{\lambda}_2)}. \tag{15}$$

Thus, the diffusion coefficient of the electrolyte depends on the value of $\bar{\lambda}$ for both ionic species.

THE ACTIVATION ENERGY

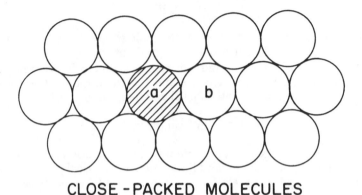

CLOSE - PACKED MOLECULES

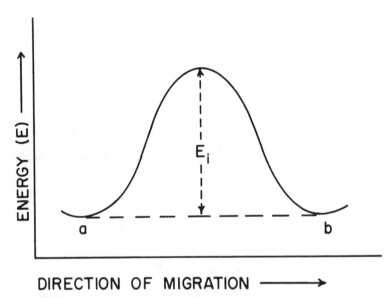

Fig. 1. Schematic representation of an ion (cross-hatched circle) surrounded by water molecules (open circles) and the energy barrier, E_i, it must surmount in moving between equilibrium positions a and b.

In the upper part of Fig. 1 an ion (cross-hatched circle) is shown surrounded by water molecules (open circles) in a close-packed arrangement. Undoubtedly, the arrangement of the ion and water molecules in a clay-water system is different than that depicted but the diagram will serve to illustrate the principle involved. If the ion moves from position a to position b it must push back the water molecules to provide space for itself at b and break bonds with the neighboring water molecules and surface charge site so that it can move into this space. Both of these processes require energy. However, the energy is released again when water molecules fill the hole that is left behind and bonds are formed between the ion and its new neighbors. The change in energy as the ion progresses from a to b is shown in the lower part of Fig. 1. Thus, for an ion to hop from one equilibrium position to the next in the direction of migration, it must pass over an energy barrier equal in height to E_i, the activation energy. Note that the hop distance is only of the order of a molecular diameter—an infinitesimal fraction of the total length of the diffusion path—and so the activation energy is independent of the geometry of this path.

According to rate process theory (6), the relation between u_i and E_i is given by

$$u_i = B_i \exp\left(-E_i/RT\right) \qquad [16]$$

in which B_i is a constant. According to the same theory, η, the coefficient of viscosity, is given by a similar equation, namely,

$$\eta = B_\eta \exp\left(E_\eta/RT\right) \qquad [17]$$

in which E_η is the activation energy for viscous flow and B_η is another constant. Note that, if $E_i \cong E_\eta$, we have

$$u_i = a_i/\eta \qquad [18]$$

where $a_i = B_i B_\eta$ and, hence, is a constant that is characteristic of the ion. Equation [18] is a form of Walden's rule which has been found to hold for Na^+ and other ions in aqueous solution (7, 24).

Recall that $\lambda_i = u_i/z_i\,F$. Therefore, we can substitute $z_i\,F\,\lambda_i$ for u_i in Eq. [18] and find

$$\lambda_i = a_i/z_i\,F\,\eta. \qquad [19]$$

It should be borne in mind that Eq. [19], like Eq. [18], applies only if the activation energy for ion movement is the same as that for viscous flow. If we disregard the difference between λ_i and $\bar{\lambda}_i$, i.e., if we disregard the variation of λ_i with distance from the particle surface, we can substitute the right-hand-side of Eq. [19] for $\bar{\lambda}_i$ in Eq. [9] to obtain

$$D_i = \beta\, b_i\, T/\eta \qquad [20]$$

in which $b_i = a_i\,R/z_i\,F$. The stokes-Einstein relation for aqueous solutions is

$$D_i = kT/6\pi r_i\,\eta \qquad [21]$$

where k is the Boltzmann constant and r is the ionic radius. Hence, we see that Eq. [20] is a form of the Stokes-Einstein relation. Also, we see that a_i and b_i can be assumed to be proportional to $1/r_i$. It should be noted here that, in keeping with Eq. [20] and [21], $D_i\eta/T$ has been found to be a constant for Na^+ and several other ions in aqueous solution (26).

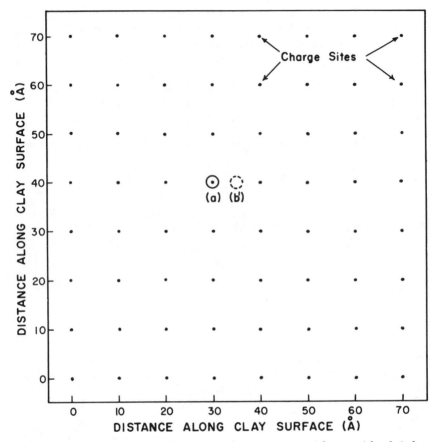

Fig. 2. Diagram showing an ion in a position of minimum potential energy (closed circle at (a)) and in a position of maximum potential energy (dashed circle at (b′)) in the electric fields of the surface charge sites (black dots).

Equations [7] and [14] indicate the factors that influence the self-diffusion of a single ionic species and the diffusion of a symmetrical electrolyte, respectively. The same factors are always operative whenever ion diffusion in clays occurs. The different factors are separately measurable except β and $\bar{\lambda}_i$. Let us attempt, therefore, to assess the magnitudes of these two quantities and determine what influences them.

Presented in Fig. 2 is a diagram showing two positions of a cation relative to the negative charge sites (black dots) on that flat surface of a clay particle. The cation is represented immediately above a charge site by the closed circle at (a) and midway between charge sites by the dashed circle at (b′). In both positions the cation is assumed to be the same distance above the plane of the charge sites. When the cation is at (a) its electrical potential energy is a minimum and when it is at (b′) its electrical potential energy is a maximum. The difference in the electrical potential energy of the cation in the two positions should equal the maximum contribution of ion-charge site interaction to the activation energy when

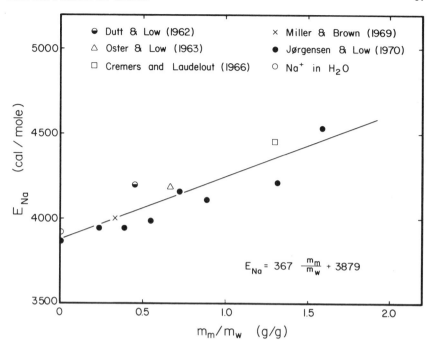

Fig. 3. Relation between E_{Na}, the activation energy for Na$^+$ conductance, and m_m/m_w, the mass ratio of montmorillonite to water.

the cation is moving parallel to the surface. Now since potential energies are additive, the electrical potential energy of the cation in each position is the summation of the electrical potential energies of the cation with respect to all the charge sites. We made the relevant summations for monovalent cations in both positions at a distance of 10 Å from the surface and found that they differ by ~40 cal/mole (14). The difference increases to ~150 cal/mole when the cations are separated from the surface by a monolayer of water and to ~1,150 cal/mole when they are in contact with the surface (23). It is likely, however, that cations in contact with the surface are essentially immobile because a combination of electrical, viscous and other forces increases their activation energy to ~9,000 cal/mole (22), which is comparable to the activation energy for ion migration in some crystalline solids (10). Since an activation energy increment of 40 to 150 cal/mole is small relative to the absolute value of the observed activation energy (e.g. see Fig. 3), it is reasonable to conclude that electrical interaction between mobile cations and charge sites contributes little to this energy. The same conclusion is reached if the clay particle is regarded as a flat condenser plate with a homogeneous charge distribution. Then, any plane parallel to the surface of the plate is an equipotential plane and a cation moving in this plane experiences no change of electrical potential. Accordingly, there is no electrical contribution to the activation energy.

Table 1. Activation energy for the conductance of exchangeable cations and the diffusion of D_2O in different homionic montmorillonites.

	Activation energy		
	Li-mont.	Na-mont.	K-mont.
		(kcal/mole)	
Exchangeable cation	4.6	4.2	4.2
D_2O	4.9	4.0	4.0

The validity of the foregoing conclusion is indicated by the data in Table 1 which were taken from the paper of Dutt and Low (4). Note that, although both the activation energy for cation conductance and the activation energy for D_2O diffusion changed with the nature of the exchangeable cation on the montmorillonite, the two activation energies were always within experimental error ($\sim \pm 0.15$ kcal/mole) of each other. Since the exchangeable cation is charged and the D_2O is uncharged, it is apparent that electrical interaction between the cation and surface charge has little effect on the energy barrier the cation must surmount in moving from one equilibrium position to another. The magnitude of the barrier depends primarily on the energy required to displace neighboring water molecules and this energy is about the same for a cation as it is for a water molecule. For the purpose of comparison, it should be noted that the activation energies for the movement of Li^+, Na^+ and K^+ in bulk water are 4,204, 3,919 and 3,435 cal/mole, respectively (12).

The same conclusion is reached by the following procedure which was reported earlier by the author (17). If B_i and B_η in Eqs. [16] and [17], respectively, have the same values in the interlayer solution as they do in a bulk solution at infinite dilution, then

$$u_i = u_i^\circ \exp[-(E_i - E_i^\circ)/RT] \qquad [22]$$

and

$$\eta = \eta^\circ \exp[(E_\eta - E_\eta^\circ)/RT] \qquad [23]$$

where the zero superscripts designate the values of the respective quantities in the bulk solution at infinite dilution. The relation between E_{Na}, the activation energy for the conductance of exchangeable Na^+, and m_m/m_w, the mass ratio of montmorillonite to water, is shown in Fig. 3 which was taken from an earlier paper (17). The equation for the best-fitting line in this figure is

$$E_{Na} = 367 \, (m_m/m_w) + 3,879. \qquad [24]$$

When $m_m/m_w = 0$, $E_{Na} = E_{Na}^\circ = 3,879$ cal/mole and so $E_{Na} - E_{Na}^\circ = 367$ m_m/m_w cal/mole. Hence, if we let i = Na in Eq. [22] and substitute 367 m_m/m_w for $E_{Na} - E_{Na}^\circ$, we find

$$u_{Na} = u_{Na}^\circ \exp[-367(m_m/m_w)/RT]. \qquad [25]$$

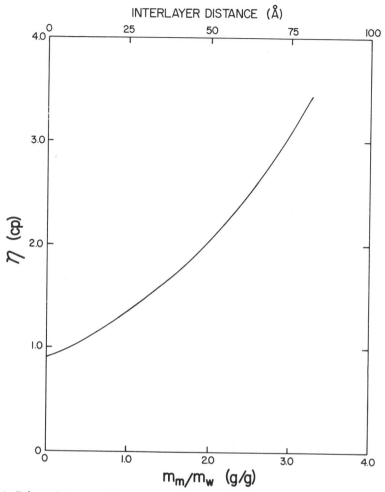

Fig. 4. Relation between η, the average coefficient of viscosity for the interlayer water, and m_m/m_w, the mass ratio of montmorillonite to water.

Also, by utilizing various methods that detect the motion of water molecules, we have found (16) that

$$\eta = \eta^\circ \exp[244(m_m/m_w)/RT].\qquad[26]$$

Comparison of Eq. [26] with Eq. [23] shows that $E_\eta - E_\eta^\circ = 244$ (m_m/m_w). Thus, we see that $E_{Na} - E_{Na}^\circ$ exceeds $E_\eta - E_\eta^\circ$ by 123 cal/mole. Since sodium ions are charged and water molecules are uncharged, this difference could be due to ion-charge site interaction. However, it is more likely due to the unequal distributions of the cations and water molecules with respect to the particle surfaces. The viscosity of the water varies with proximity to these surfaces as illustrated by the curve in Fig. 4, which is described by Eq. [26] with $\eta^\circ = 0.8937$ cp and $T = 298.16$ K. If the cations and water molecules had the same distribution, this variation in viscosity would have the same effect on their activation energies. However, the electric fields of the particles affect the two species differently and the concentration of cations increases exponentially with proximity to the particle surfaces, whereas, the concentration of water molecules remains virtually constant. The net result is that a larger fraction of the cations are moving in water of enhanced viscosity (where the energy barriers for the displacement of water molecules are higher) and so $E_{Na} - E_{Na}^\circ > E_\eta - E_\eta^\circ$. In the development of the foregoing equations, we ignored the possibility that the activation energy might vary with the distance from the particle surfaces in order to avoid complications that have been noted elsewhere (9).

Both theoretical arguments and experimental results support the conclusion that, at any distance from the surface of a particle, the activation energy for ion movement in clay-water systems is essentially the same as that for viscous flow, at least for monovalent ions. This means that Eqs. [18] and [19] apply to the interlayer regions in such systems and that the viscosity of the interlayer water is an important factor governing ion movement.

For the sake of convenience, let us disregard the difference between λ_i and $\bar{\lambda}_i$ and between u_i and $\bar{u}_i$. Then we can combine Eqs. [20] and [26] to obtain

$$D_i = \frac{\beta b_i T}{\eta^\circ} \exp\left[-244(m_m/m_w)/RT\right] = \beta D_i^\circ \exp\left[-244(m_m/m_w)/RT\right] \qquad [27]$$

where D_i° is the diffusion coefficient in bulk water. Also, we can combine Eqs. [11] and [25] to obtain

$$U_{Na} = \beta u_{Na}^\circ \exp\left[-367(m_m/m_w)/RT\right]. \qquad [28]$$

Consequently, if we have measured values of D_i at various values of m_m/m_w and know the value of D_i°, we can use Eq. [27] to calculate the corresponding values of β. Or, alternatively, if we have measured values of U_{Na} at various values of m_m/m_w and know the value of u_{Na}°, we can make the same calculations.

Shown in Fig. 5 are the combined data from three different studies (11, 21, 25) on the dependence of D_{Na} in Na-saturated, Upton montmorillonite on m_m/m_w at 25 C. Shown in Fig. 6 are the combined data from two different studies (2, 9) on the dependence of U_{Na} on m_m/m_w in the same montmorillonite and at the same temperature. In the latter studies, U_{Na} was not reported but we calculated it from the data given by using an equation developed earlier (20), viz.,

$$L = z_i U_i c_i F/1{,}000 \qquad [29]$$

in which L is the specific conductance. The value of D_{Na}° is 1.32×10^{-5} cm²/sec (27) and the value of u_{Na}° is 5.19×10^{-4} cm²/volt sec. This value of

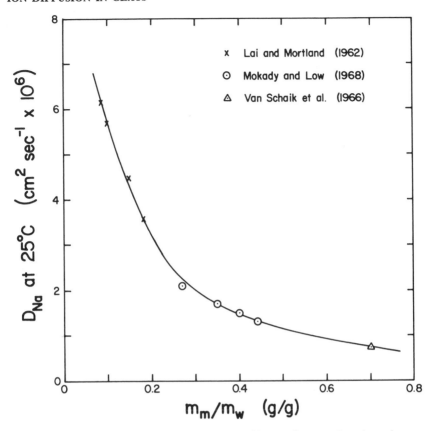

Fig. 5. Relation between D_{Na}, the self-diffusion coefficient of Na$^+$, and m_m/m_w, the mass ratio of montmorillonite to water.

u_{Na}° was calculated from the limiting equivalent conductivity of Na$^+$ reported by Stokes and Robinson (24). Thus, we were able to determine β as a function of m_m/m_w by two different methods, namely: 1) by using the data in Fig. 5 and Eq. [27] and 2) by using the data in Fig. 6 and Eq. [28]. We shall refer to these methods as Method I and Method II, respectively. The results obtained by using them are presented in Fig. 7. Also presented in this figure are results calculated by means of the Burger-Fricke equation, discussed by Cremers et al. (3), which can be written

$$\beta = \phi/[\phi + k(1-\phi)] \qquad [30]$$

where ϕ is the porosity and k is a constant governed by the axial ratio of the clay.

Note from Fig. 7 that the values of β calculated by Method I do not agree with those calculated by Method II. We do not know the reason for the disagreement. It may be that the values of U_{Na} in Fig. 6 are high because of the electroosmotic flow of water in the clay-water system when an electrical potential gradient is imposed. Or it may be that this gradient

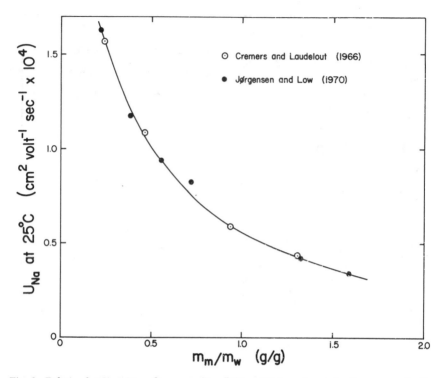

Fig. 6. Relation between U_{Na}, the apparent mobility of Na⁺, and m_m/m_w, the mass ratio of montmorillonite to water.

tends to orient the particles with their long axes parallel to the direction of ion movement and, thereby, increase β.

Also, note from Fig. 7 that the values of β calculated by means of Eq. [30] agree reasonably well with those calculated by Method I if an appropriate value of k (i.e., 38.9) is chosen, but do not agree with those calculated by Method II even if an appropriate value of k (i.e., 13.9) is chosen. The latter value of k is the one that Cremers et al. (3) found to be suitable. In view of the fact that Method I and Eq. [30] give essentially the same results, we are inclined to regard Method I as being more reliable than Method II. Therefore, we believe that the open circles in Fig. 7 represent the dependence of β on m_m/m_w for ion diffusion in montmorillonite-water systems of low electrolyte content.

Finally, let us return to the concept that, except in self-diffusion, an ionic species does not move independently of the other ionic species that are present. If one ionic species moves faster than another, a separation of charge occurs and an electrical potential gradient is established which retards the faster moving species and accelerates the slower moving one so that electroneutrality is maintained in any macroscopic region. This principle is embodied in Eq. [14] and is illustrated by the data in Table 2, which were taken from the work of Husted and Low (8). To obtain these data, a 0.1 N KCl solution was separated from a 0.1 N chloride solution of another monovalent cation, M⁺, by a montmorillonite plug and the

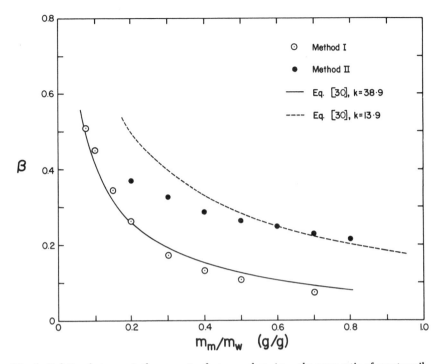

Fig. 7. Relation between β, the geometry factor, and m_m/m_w, the mass ratio of montmorillonite to water.

amounts of K^+ and M^+ diffusing in opposite directions through the plug were determined daily. The mobilities of the relevant ions in aqueous solution at infinite dilution are in the order $H > K = NH_4 > Na > Li$ (24). Observe that the rate of K^+ diffusion depended on the counterdiffusing ion and decreased as the mobility of this ion decreased. Fortunately, H^+ diffuses out of and away from plant roots and, since it is the most mobile of all cations, the counterdiffusion of nutrient ions toward the roots is facilitated.

Table 2. Counterdiffusion of K^+ and another monovalent cation, M^+, through montmorillonite.

System	Daily diffusion		
	K^+	M^+	Cl^-
		(meq)	
K-H	0.026†	--	--
K-NH$_4$	0.016	0.017	−0.001‡
K-Na	0.016	0.013	+0.003
K-Li	0.013	0.008	+0.005

† L.S.D. is 0.0004 or less for any two values within any column or from one colume to another.
‡ A positive value indicates that Cl^- diffused with K^+.

In summary, we have shown that the diffusion of any ion in a clay-water system is influenced primarily by the geometry of the diffusion path as governed by the arrangement of the clay particles, the viscosity of the water between the particles and the mobility of the codiffusing or counterdiffusing ion. Also, we have presented data that allow the effects of these factors to be determined separately for a system of Na-montmorillonite and water at any ratio of montmorillonite to water.

LITERATURE CITED

1. Banin, A., and P. F. Low. 1971. Simultaneous transport of water and salt through clays: 2. Steady-state distribution of pressure and applicability of irreversible thermodynamics. Soil Sci. 112:69–88.

2. Cremers, A. E., and H. Laudelout. 1966. Surface mobilities of cations in clays. Soil Sci. Soc. Am. Proc. 30:570–576.

3. ————, J. van Loon and H. Laudelout. 1966. Geometry effects for specific electrical conductance in clays and soils. In S. W. Bailey (ed.) Clays and clay minerals, Proc. 14th Natl. Conf., Berkeley, CA. p. 149–162.

4. Dutt, G. R., and P. F. Low. 1962. Relationship between the activation energies for deuterium oxide diffusion and exchangeable ion conductance in clay systems. Soil Sci. 93:195–203.

5. ————. 1962. Diffusion of alkali chlorides in clay-water systems. Soil Sci. 93:233–240.

6. Glasstone, S., K. J. Laidler, and H. Eyring. 1941. The theory of rate processes. McGraw-Hill Book Co., New York.

7. Harned, H. S., and B. B. Owen. 1943. The physical chemistry of electrolytic solutions. Reinhold Publishing Corp., New York.

8. Husted, R. F., and P. F. Low. 1954. Ion diffusion in bentonite. Soil Sci. 77:343–353.

9. Jørgensen, P., and P. F. Low. 1970. Conductance studies of Na-bentonite-water systems. Isr. J. Chem. 8:325–333.

10. Jost, W. 1952. Diffusion in solids, liquids, gases. Academic Press, New York.

11. Lai, T. M., and M. M. Mortland. 1962. Self-diffusion of exchangeable cations in bentonite. In E. Ingerson (ed.) Clays and clay minerals, Proc. 9th Natl. Conf., West Lafayette, IN. p. 229–247.

12. Low, P. F. 1958. The apparent mobilities of exchangeable alkali metal cations in bentonite-water systems. Soil Sci. Soc. Am. Proc. 22:395–398.

13. ————. 1960. Influence of adsorbed water on electrochemistry of clays. Int. Congr. Soil Sci. Trans. 7th, Madison, WI. II:328–336.

14. ————. 1962. Influence of adsorbed water on exchangeable ion movement. Clays Clay Miner. 9:219–228.

15. ————. 1968. Observations on activity and diffusion coefficients in Na-montmorillonite. Isr. J. Chem. 6:325–336.

16. ————. 1976. Viscosity of interlayer water in montmorillonites. Soil Sci. Soc. Am. Jour. 40:500–505.

17. ————. 1976. Relation between the viscosity and other properties of water in montmorillonites. I:24–30. In M. Kutilek and J Suter (ed.) Proc. symposium on water in heavy soils. Bratislava, Czechoslovakia.

18. ————, and G. R. Dutt. 1964. Diffusion of alkali-chlorides in clay-water systems. II. Response to Bolt-de Hahn comment. Soil Sci. 97:346–349.

19. Miller, R. J., and D. S. Brown. 1969. The effect of exchangeable aluminum on conductance and diffusion activation energies in montmorillonite. Soil Sci. Soc. Am. Proc. 33:373–378.

20. Mokady, R. S., and P. F. Low. 1966. Electrochemical determination of diffusion coefficients in clay-water systems. Soil Sci. Soc. Am. Proc. 30:438–442.

21. ————. 1968. Simultaneous transport of water and salt through clays: I. Transport mechanisms. Soil Sci. 105:112–131.

22. Oster, J. D., and P. F. Low. 1963. Activation energy for ion movement in thin water films on montmorillonite. Soil Sci. Soc. Am. Proc. 27:369–373.

23. Shainberg, I., and W. D. Kemper. 1966. Conductance of adsorbed alkali cations in aqueous and alcoholic bentonite pastes. Soil Sci. Soc. Am. Proc. 30:700–706.

24. Stokes, R. H., and R. A. Robinson. 1955. Electrolyte solutions. Academic Press, New York.

25. van Schaik, J. C., W. D. Kemper, and S. R. Olson. 1966. Contribution of adsorbed cations to diffusion in clay-water systems. Soil Sci. Soc. Am. Proc. 30:17–22.

26. Wang, J. H. 1954. Effect of ions on the self-diffusion and structure of water in aqueous electrolyte solutions. J. Phys. Chem. 58:686–692.

27. ————, and S. Miller. 1952. Tracer diffusion in liquids. II. The self-diffusion as sodium ion in aqueous sodium chloride solutions. J. Am. Chem. Soc. 74:1611–1612.

CHAPTER 4

Charge Properties in Relation to Sorption and Desorption of Selected Cations and Anions

A. MEHLICH[1]

[1] Agronomic Division, North Carolina Dep. of Agric., Raleigh, NC 27611.

ABSTRACT

Charge properties of soil have a vast impact on the internal soil environment. These properties include constant negative charge cation exchange capacity (CECc), variable charge cation exchange capacity (CECv), total negative charge (CECt), positive charge or anion exchange capacity (AEC), and anion sorption capacity (ASC). A high proportion of CECc in relation to CECt corresponds to 3-layer phyllosilicate minerals, high Si to Al ratios (greater than 2:1), and unbuffered salt exchangeable H^+/Al^{3+} of organic acids. High proportions of CECv are associated with 2-layer phyllosilicate minerals, low Si to Al ratios (less than 2:1), and amorphous clays and organic acids, where acidity due to H-bonding is not exchangeable by a cation of unbuffered salt, and where neutralization by base leads to the development of variable charge. Positive charge and anion sorption capacity originate from the protonation of sesquihydroxides or their hydrates. Positive charge is produced through protonation of hydroxyl from sesquihydroxide surface countered by monovalent anions, such as Cl^- or NO_3^-. The quantity of Cl or NO_3 desorbed by another anion is suggested to be equivalent to AEC. The specific site of positive charge is an aquo ligand. An aquo ligand when exchanged by an oxo ligand results in the deletion of positive charge. Similarly, if protonation by acids having tetrahedral structure, such as H_2SO_4 or H_3PO_4, occurs positive charge fails to develop. Oxo ligands are not exchangeable by H ions or acidic hydrolizable anions, but they are exchangeable by hydroxy anions or basic ionizable anions. The quantity of hydroxyl sorbed in this reaction may be recorded as anion sorption capacity (ASC). The sum of specifically identified desorbed anions in this reaction also represent ASC.

Sorption of cations where CECc predominate (smectites) increases in the order Li < Na < K < H < Mg < Ca, while for ground muscovite the order is Li < Na < Mg < Ca < K < H. Where CECv is predominant (kaolinite) sorption is in the order Li < Na < H < K < Mg = Ca, while sorption by H-humus at low solution concentration is Li < Na < K < Mg < Ca < Sr < Ba < La. Cation bonding energies by layer minerals are on the average twice as high for divalent than monovalent cations.

Sorption of phosphoric acid and its potassium salts increased and desorption of sulfate decreased in the order K_3PO_4, K_2HPO_4, KH_2PO_4, and H_3PO_4. Due to its basic hydrolysis ($PO_4^{3-} + H_2O \rightleftharpoons HPO_4^{2-} + OH^-$), K_3PO_4 was most effective in desorption of S, increasing soil pH, and decreasing negative charge soil acidity. In contrast, H_3PO_4 due to its primary ionization ($H_3PO_4 \rightleftharpoons H^+ + H_2PO_4^-$) was ineffective in desorption of S, in decreased soil pH, and increased soil acidity. K_2HPO_4 reacted intermediate to H_3PO_4 with respect to the above properties. Desorption of S and P was primarily a function of hydroxyl ion concentration with S appearing in solution as SO_4^{2-} and P appearing in solution primarily as HPO_4^{2-}.

In addition to K_3PO_4 and K_2HPO_4, desorption of S was equally effective with Na_2SiO_3 and NH_4VO_3. This was followed with $KHPO_4$, KOH,

$(NH_4)_6$ Mo_7O_{24}, EDTA and NH_4F; lower desorption of S was with KH_2PO_4, $NaBO_2$, $(HPO_3)_X$ and desorption was least with H_3PO_4 and CH_3COONa. Reasons for these differences are based on neutralization capacity of anion for exchange and variable charge acidity, special anion and chelation effects.

Future progress in the evaluation of charge characteristics of soils and minerals requires methodology which identifies both cation and anion charge characteristics involved in the reaction.

KIND AND QUANTITY OF NEGATIVE AND POSITIVE CHARGE

Negative Charge

Kind and quantity of negative and positive charge have great scientific and practical importance in soil-plant nutrition relationships; sorption and de-sorption of ions; soil classification; water and physicochemical relationships; and transport and transformation of pesticide, animal, municipal, and industrial waste to soil. Sources of negative charge are the layer silicates, layer silicate-sesquioxide complexes, and organic matter; while presence of sesquioxide hydrates are primarily responsible for positive charge. A foremost contribution to kind of negative charge was made by Schofield in 1939, when soils were acid penetrated, saturated with the cation of a neutral salt and separated into "permanent" and pH-dependent charge. Schofield suggests that the precise determination of these variables as a function of pH encounters difficulties and he further concludes that a fixed pH "is not in itself sufficient to fix the charge". The concepts relating to "permanent" and pH-dependent charge were further developed by Schofield (1949) and by Fields and Schofield (1960); the term "permanent" charge was retained while "variable" was substituted for pH-dependent charge.

Schofield in 1939 showed that bentonite leached with dilute HCl and washed free of excess acid, reacted with about the same quantity of alkali in reaching the neutral point as an equivalent amount of concentrated KCl extractable acidity at pH 1.5. Accordingly, acid-washed bentonite appeared to entirely exchange acidity which in effect carried constant negative charge within pH range 1.5 to 6.0. This constant charge was believed to be due to isomorphous replacements in the crystal lattices. Coleman and Thomas (1967) also concluded that this concept is perfectly valid when dealing with moderately pure layer silicates, but that it cannot be extended "to soils, where organic matter, intergrade minerals, and perhaps allophane are present". These reasons and other evidence involving CEC and pH by Pratt and Bair (1962); Helling, Chesters, and Corey (1964); Rich and Obenshain (1955); Ross, Lawton, and Ellis (1964); Thomas (1960); Low (1955); Bhumbla and McLean (1965); Coleman, Weed, and McCracken (1959); Jenny (1961); Paver and Marshall (1934); McLean, Reicosky, and Lakshmanan (1965); and others prompted a SSSA subcommittee on soil chemistry terminology (Glossary, 1979) to list the term "permanent charge" obsolete and redefine it "the net negative (or positive) charge of clay particles inherent in the crystal lattice of the particle; not affected by changes in pH or by ion-exchange reactions."

Cation or anion exchange capacity which varies with pH has been defined by Schofield (1939) as pH dependent charge and by Fieldes and Schofield (1960) as variable charge where at high pH Schofield attributed negative charge to the ionization of H from SiOH groups. This source of charge has been largely discounted by Goates and Anderson (1956) who showed that SiOH groups do not appreciably ionize in the pH range concerned. Major contributions to variable charge are the layer silicate-sesquioxide complexes (Coleman and Thomas, 1964; Schwertmann and Jackson, 1964; Volk and Jackson, 1964). Other sources of variable charge include allophanes, aluminous pedogenic chlorite (deVilliers and Jackson, 1967; Birrell, 1958; Birrell and Gradwell, 1956; Wada and Ataka, 1958; Aomine and Jackson, 1959) and organic matter. Negative charge of the latter arise from such acid functional groups as carboxyls, phenols, enols, and possibly alcoholic hydroxyls (Gillam, 1940; Broadbent and Bradford, 1952; Lewis and Broadbent, 1961; Schnitzer and Desjardins, 1962; Schnitzer and Gupta, 1964, 1965; Schnitzer and Skinner, 1963; Schnitzer and Wright, 1960).

Positive Charge

Positive charge originates from protonated sesquioxidehydrates and has been measured by Schofield (1939, 1949) via uptake of Cl^- from alcoholic HCl. Chloride anion sorption occurs only below the isoelectric point of the reactive groups. Other investigators who used sorption of chloride to determine the size and magnitude of electrical charges carried by clay, oxide mineral, and allophanic tropical soils were, Quirk, 1960; Sumner and Reeve, 1966; and Gebhardt and Coleman, 1974a. Chloride sorption capacities of tropical soils (principally Dystrandepts) were determined by Gebhardt and Coleman by shaking 5 g air dry samples of soil for 1 hour in 20 ml 0.17N $AlCl_3$, measuring Al and Cl in the filtrate, and calculating sorbed Al and Cl by difference. Sorption of Cl was concentration-dependent and essentially consistent with the Langmuir adsorption equation. The authors concluded that the data were compatable with the reaction mechanism of Hingston et al. (1967). Accordingly, positive sites for "nonspecific absorption" are created through the acceptance of protons by octahedrally coordinated Al^{3+} or Fe^{3+}.

While sorption of chloride (or nitrate) of protonated soil appears to be an acceptable measure of positive charge, sorption of sulfate and phosphate by a "ligand exchange reaction" as defined by Hingston et al. (1967) is not. When data on the same soils were expressed in mmols/100 g Gebhardt and Coleman (1974a, 1974b, 1974c) reported approximately maximum sorption of 27, 30, and 38 mmols of chloride, sulfate, and phosphate, respectively. The corresponding data expressed on an equivalent basis would be 27, 60, and 114 meq/100 g soil.

Methodology

In view of the wide range of concepts regarding sorption and desorption mechanisms of cations and anions by soils and minerals, it is important to adequately characterize charge properties of soils consistent with existing knowledge. Methodology was initially developed to separate cation and anion exchange capacity by Mehlich (Mehlich, 1953). Procedures for separating total negative charge into "permanent" and pH-dependent or variable cation exchange capacity was developed by Schofield (1939, 1949); Fieldes and Schofield (1960); Coleman, Weed, and McCracken (1959); and Mehlich (1960). For calculating permanent charge cation exchange capacity, Coleman et al. used the sum of KCl extractable H^+, Al^{3+}, Ca^{2+}, Mg^{2+}, and NH_4OAc extractable K^+ and designated the results as "effective" charge (ECEC). The quantity of residual acidity, measured with $BaCl_2$-triethanolamine, which remained after neutral salt leaching, was recorded as pH-dependent charge. Data recorded as ECEC increased with increasing pH of the natural soil and does not represent permanent charge except where unbuffered salt exchangeable H^+ plus Al^{3+} exceeds the sum of K^+, Ca^{2+}, Mg^{2+}, Na^+.

Since total soil acidity can be selectively classified into unbuffered salt-exchangeable or replacable and residual acidity, it would be a logical consequence to prepare a completely acid saturated soil and to separate the varying quantities of charge components as outlined by Mehlich in 1960. Following determination of unbuffered salt exchangeable H^+ and Al^{3+} with 0.6N $BaCl_2$ and determination of residual acidity with $BaCl_2$-TEA, total charge at pH 8.2 was determined with Ba as the "index" cation following replacement with 0.6N $CaCl_2$. A separate sample was pretreated with HCl, then saturated with $BaCl_2$ and the quantity of the Ba replaced with $CaCl_2$ was designated as "permanent" charge cation exchange capacity. This portion of the procedure was modified by Mehlich (Characterization of Negative Charge Properties (CEC) Based on Hydrogen and/or Aluminum Saturated Soils, Agronomy Abstracts, 1975 Annual Meetings, p. 179) by pretreating the soil with HCl followed by equilibration with 0.1N HCl-$AlCl_3$ and displacing H^+/Al^{3+} with 0.6N $BaCl_2$. Exchangeable H^+ and Al^{3+} were recorded separately and soils classified on the basis of $Al^{3+}/Al^{3+} + H^+$ ratios while the sum of H^+ and Al^{3+} were designated as unbuffered salt exchangeable cation exchange capacity (CECc).

Residual acidity and total negative charge (CECt) were determined with $BaCl_2$-TEA as before (Mehlich, 1960) and the difference between CECt and CECc was designated as variable charge cation exchange capacity (CECv).

While Coleman et al. (1959) designated the $BaCl_2$-TEA titratable acidity, after salt extractable acidity, as pH-dependent charge, Mehlich (1960) separated this form of acidity into acidity due to variable negative charge and due to neutralizable anion exchange acidity. The need for this separation was further substantiated by the results obtained on the sorption and desorption of sulfate (Mehlich, 1964a, 1964b). Triethanolamine $(CH_2CH_2OH)_3N$ reacts simultaneously to neutralize variable charge acidity and to supply hydroxyl in exchange for sorbed anions. In the presence of $BaCl_2$, Ba^{2+} ions are sorbed in exchangeable form by CECv sites while Ba reacts with the hydroxyl-substituted anion species to form Ba salts. These two forms of acidity were designated by Mehlich (1960) as variable charge acidity (Hv) and neutralizable anion exchange acidity (Han). In the case of acid pretreated soil, Hv plus CECc should be equal to CECt if the charged sites of soils are negative, but if the charged sites of soils are also positive or contain hydroxyl-exchangeable anions, Hv plus CECc would be expected to exceed CECt with the difference being designated as Han. In terms of quantity of positive charge or anion exchange, Han would be equivalent to anion exchange capacity (AEC), or anion sorption capacity (ASC) as subsequently defined.

CHARGE PROPERTIES OF REPRESENTATIVE SOILS AND MINERALS

Representative charge properties of layer silicate soils and minerals, mixed crystalline and amorphous constituents, and histosols are presented in Tables 1 to 3. The data include CECc, CECv, CECt, and ASC as outlined above, with the results expressed in meq/100 g soil or mineral. The unit ASC (anion sorption capacity) rather than AEC (anion exchange capacity) has been chosen since this determination involved the "sorbed" phosphate anion. The charge properties in these tables have been presented also in terms of percent CECt under charge distribution.

The clay constituents of the soils listed under Kenya were determined by x-ray and DTA analysis according to Theisen (1966a) and those listed under NC (North Carolina) were determined by Sterling B. Weed, North Carolina State University.

Layer Silicates

Charge characteristics and charge distribution of layer silicate minerals and soils are recorded in Table 1. White Store is the B horizon of Triassic silt-stone, Piedmont, North Carolina (NC), and Georgeville is the B horizon derived from methamorphosed felsic rock, Piedmont, NC. The soils from Kenya were Songhor B horizon black clay loam, South Nyanza Province, Tavete; B1 horizon of dark red volcanic loam, Southern Province; and Madungoni B horizon of red sand, Coast Province.

Where montmorillonite (M) and vermiculite (V) predominated, CECc comprised 85 to 100% of CECt, while CECv was proportionately low and ASC was also low. In the presence of hydrous mica, CECc was within 60 to 65%, CECv between 35 to 40%, and ASC between 11 to 14% of CECc. In the case of Mogutato soil where V/M was predominant CECc was 59, CECv 41, and ASC 36% of CECt. Higher ASC was due to the presence of 20 to 50% kaolinite (K). With K and with soils where K predominated CECc comprised 25 to 52, CECv 48 to 75, and ASC 38 to 54% of CECt. Highest ASC of 84% was with halloysite (H).

Sorption of cations from neutral chloride salts by soils showed high sorption of cations and negligible sorption of chloride anion (Mattson,

Table 1. Charge properties of layer silicates.

Soils or minerals	Crystalline constituents†	Charge properties (see text)				Charge distribution		
		meq/100 g				% of CECt		
		CECc	CECv	CECt	ASC	CECc	CECv	ASC
Bentonite, Athy River, Kenya	M_1	112.0	6.0	118.0	1.0	95	5	1
Vermiculite, S. Africa	V_1	85.0	0.0	85.0	0.0	100	0	0
Songhor, Kenya	$M_1(IKH)_2$	44.2	3.4	47.6	3.5	93	7	7
White Store, NC	M_1K_3	17.2	3.0	20.2	1.1	85	15	5
Hydrous mica, Ill	I_1	11.5	7.7	19.2	2.7	60	40	14
Taveta, Kenya	M_1I_2	24.4	13.0	37.4	4.1	65	35	11
Mogutato, Kenya	$V/M_1K_2I_3$	17.0	12.0	29.0	10.4	59	41	36
Kamec Kaolin, NC	K_1	3.2	4.2	7.4	4.0	43	57	54
Kaolinite	K_1	1.1	3.3	4.4	2.0	25	75	45
Halloysite	H_1	5.5	12.3	17.8	14.9	31	69	84
Madungoni, Kenya	K_1	3.0	2.8	5.8	2.2	52	48	38
Georgeville, NC	$K_1(VC)_2$	7.2	7.6	14.8	6.4	49	51	43

† M—montmorillonite; V—vermiculite; I—hydrous mica; K—kaolinite; H—halloysite; C—chlorite. Approximate quantities: 1, >50; 2, 20–50; 3, <20.

Table 2. Charge properties of mixed crystalline and amorphous constituents.

Soils or minerals	Crystalline†	Amorphous‡	Charge properties (see text)				Charge distribution		
			CECc	CECv	CECt	ASC	CECc	CECv	ASC
			meq/100 g				% of CECt		
Al-tetrasilicate		XXX	48.0	2.0	50.0	--	96	4	--
Al-metasilicate		XXX	19.6	88.4	108.0	--	18	82	--
Allophane (Egmont)		XXX	10.3	40.7	51.0	17.0	20	80	33
Gibbsite	Gi_1	X	0.0	5.5	5.5	5.5	0	100	100
Goethite	Go_1	X	0.0	4.1	4.1	4.0	0	100	98
Cecil, B_3, NC	$K_1V_2(GiGo)_3$	X	3.4	3.8	7.2	6.0	47	53	83
Hayesville, B_{22}, NC	K_1Gi_2	X	1.0	5.0	6.0	6.4	17	83	106
Kanja, B, Kenya	KH_2 (V/M Gi)$_3$	XX	7.2	24.2	31.4	13.1	23	77	42
Gatare, B, Kenya	(V/M/C Gi)$_2$	XXX	2.0	24.0	26.0	18.4	8	92	71
Kamwete, B, Kenya	(I/C KH Gi)$_2$	XXX	1.7	13.7	15.4	16.2	11	89	105

† See footnote, Table 1.
‡ Amorphous components: XXX = dominant; XX = detected, X = not detected.

1928, 1931). Sorption was in the order Na < K < Ca < Ba, using a Sharkey clay. Schachtschabel (1940) found relative ion affinities, measured by adding one symmetry concentration of alkali or alkali earth chlorides to NH_4 clay to be:

Montmorillonite Li < Na < K < H < Mg < Ca
Kaolinite   Li < Na < H < K < Mg = Ca
Ground muscovite Li < Na < Mg < Ca < K < H

Such series as listed above are found to vary somewhat with the concentration of the salt solution and with the ion initially adsorbed on the clay (Jenny, 1932; Schachtschabel, 1940; Vanselow, 1932), but differences between types of clay are evident, regardless of such effects.

Cation bonding energies by layer minerals were reported by Marshall (1964) on an average to be twice as high for divalent than monovalent cations on a molar basis. Calculations based on data by Chatterjee (1956) and Chatterjee and Marshall (1950) resulted in peaks and valleys of bonding energies in relation to percent of equivalence. Cation bonding energies greatly diminished at and above percent equivalence. When two or more cations were present on the clay simultaneously, free energy relationships followed the rule of Jarusov (1937). The bonding energy of divalent cations for kaolinite was Mg > Ca > Ba, for two bentonites it was Ba > Ca and for Putnam clay (mollic albaqualf) the order was inconsistent and varied slightly with meq of base added.

Exchange isotherms for the entry of six cations into NH_4-Putnam clay, against symmetry amounts of electrolyte added were reported by Gieseking and Jenny (1936). Sorption of cations increased with increasing symmetry concentration up to four and in the order Ba > Ca > Mg > K > Na = Li. Gieseking (1939, 1949) also showed that clay minerals enter into exchange reactions with organic cations. These cations are ammonium ions in which one or more of the hydrogens have been substituted by organic groups. They are strongly sorbed by montmorillonic clay minerals and difficult to replace by metal cations but more easily desorbed by other large organic cations.

For a detailed account on the historical developments of cation exchange and use of cation exchange equations the reader is directed to the eloquent review by G. W. Thomas (1977).

Mixed Crystalline and Amorphous Constituents

Representative data on charge properties of selected minerals, mixed crystalline, amorphous constituents, and soils are shown in Table 2. Al tetra and meta silicates are laboratory preparations as described by Mehlich (1967), while allophane (Egmont) was obtained from New Zealand (Fieldes, 1955). Gibbsite was a synthetic preparation by decomposing potassium aluminate with CO_2 and the Goethite mineral was from Ishpeming, Michigan. Cecil was a B3 horizon sample derived from siliceous crystalline rock, Piedmont, NC, while Hayesville B22 was from hornblend gneiss, Piedmont, NC. The three soils from Kenya were Kanja (0 to 15 cm) red brown clay loam, Central Province; Gatare A horizon derived from volcanic ash, in the Aberderes, Central Province; and Kamweti A horizon derived from volcanic ash, Mt Kenya, Central Province. Differential thermal analysis curves of the < 2M A horizon clays saturated with Mg had a primarily endothermic effect between 100 and 250 C with both the Gatare and Kamweti soils, although the latter showed an endothermic effect at about 350 C (Theisen, 1966b). This effect indicated the presence of gibbsite which was dissolved by treatment with warm 0.5N NaOH.

Amorphous alumino silicates prepared by reacting $AlCl_3$, Na_2Si_4O (tetra silicate), and NaOH in the presence of either $MgCl_2$, $CaCl_2$, or $BaCl_2$ exhibited 96% of CECt as CECc, while those prepared from $AlCl_3$, Na_2SiO_3 (meta silicate), and NaOH exhibited 18% CECc and 82% CECv. Allophane from New Zealand had predominantly CECv and 33% ASC. Gibbsite and goethite had 100% CECv and 98 to 100% ASC. Charge distribution of Cecil was 47, 53, and 83% CECc, CECv, and ASC, respectively. The soils containing gibbsite, goethite, and in combination with amorphous components had high CECv while ASC ranged between 42 to 105% of CECt.

Histosols

Charge properties and charge distribution of organic soils are reported in Table 3. Humic acid was a Merck and Co, Inc. commercial preparation. Peat was from Florida and Typic medasaprist from the Tidewater area, North Carolina. Moorland and Kamveti were surface samples of histosols of the Aberdare and Mt Kenya plateaus, Kenya, respectively.

Charge distribution of the organic materials were predominantly CECv while ASC was low except in the Kamwete sample, which was observed to contain amorphous ash, where CECt was low in comparison with the other organic soils.

Charge properties of histosols are greatly influenced by kind and quantity of mineral components. For example, Dolman and Buol (1967) observed that with high organic matter (about 75%), acid histosols contained more KCl extractable hydrogen ions than aluminum ions whereas with increasing silt and clay content KCl extractable Al^{3+} in proportion to H^+ increased. Mehlich and Bowling, 1975 (Exchange Characteristics of Aluminum and Hydrogen Medisaprists and an Aluminum Hapludalf, Agronomy Abstracts, 1975 Annual Meetings, p. 176) obtained similar results by extraction with 0.6N $BaCl_2$, except that the recovery of Al^{3+} and H^+ was considerably higher and decreased in the order; Ba > Ca > K = NH_4 > Mg > Na. Histosols equilibrated with equal concentrations of an H^+/Al^{3+} mixture (0.1N $AlCl_3$ in 0.1N HCl) were also observed to selectively remove from this mixture increasing proportions of H^+ over Al^{3+} as the organic matter increased while the clay and silt contents decreased. This method has been described as a possible supplementary laboratory index in the taxonomic classification of Medisaprists and mineral soils having a Histic epipedon on the family level.

Table 3. Charge properties of histosols.

Soils	Charge properties (see text)				Charge distribution		
Soils	CECc	CECv	CECt	ASC	CECc	CECv	ASC
	meq/100 g				% of CECt		
Humic acid	62.0	208.0	270.0	0.0	23	77	0
Peat, Florida	38.0	98.0	136.0	5.5	28	72	4
Moorland, Ap, Kenya	33.5	86.5	120.0	10.0	28	72	8
Kamwete, Ap, Kenya	5.0	18.0	23.0	4.6	22	78	20
Dare, Ap, NC (Typic medisaprist)	60.0	18.0	240.0	1.1	25	75	< 1

INFLUENCE OF CATIONS AND ANIONS ON CHARGE PROPERTIES

Mineral Soils and Synthetic Aluminum Silicates

In the case of crystalline layer silicate minerals, excess negative charge is known to be developed by isomorphous substitution. Excess negative charge (CECc) was also indicated with synthetic aluminum silicates as shown in Table 2. Mattson (1928, 1931) reported charge data from bulk preparations of metasilicate (acidoid) and aluminum salts (basoid) without being able to distinguish between kind of charge. Similar synthetic preparations from $AlCl_3$ and Na_2SiO_3 (sodium meta silicates) by Mehlich (1967) showed the charge to be primarily CECv. However, when $AlCl_3$ was reacted with $Na_2Si_4O_9$ (sodium tetrasilicate), and following 10 days of aging, the charge was primarily CECc (CECp listed in the publication). Charge capacities increased greatly with increasing Al to Si atomic ratio between 0.5 to 1.5 with the $AlCl_3$-Na_2SiO_3 combination and between 1 to 5 with the $AlCl_3$-$Na_2Si_4O_9$ reaction. Simultaneous mixing of $AlCl_3$ and $Na_2Si_4O_9$ with adjustment to pH 5.0, 7.0, and 8.3 produced a separable precipitate only in the presence of normal solutions of either $CaCl_2$, $MgCl_2$, or $BaCl_2$. Under these conditions, CECc increased with increasing Al to Si ratio to 1:5 at pH 5.0 nd to 1:3.5 at pH's 7.0 and 8.3. At pH 5.0 and 7.0, CECv exceeded CECc when the Al to Si ratios were less than 3.5 and 1.75, respectively, while at pH 8.3 CECc exceeded CECv at all Al to Si ratios above 1. When either $Na_2Si_4O_9$ or $H_2Si_4O_9$ was reacted with $Al(OH)_3$-gel the measured charge properties were exclusively CECv.

It appears therefore that in addition to isomorphous substitution, constant or invariable charge (CECc) may be measured when the Al to Si ratios are high, when the silicate structure is not too condensed, when the base to Si ratios are narrow (viz. 0.5 for $Na_2Si_4O_9$), and when synthesis proceeds in the presence of divalent cations. In contrast, variable charge (CECv) develops when Al to Si ratios are low, when the silicate structure is condensed, and the base to Si ratios are high (viz. 2.0 for Na_2SiO_3).

An explanation for these differences in charge properties were offered by Mehlich (1967) on the basis of the Hofmann, Endell, and Wilm (1933) structures of crystalline layer silicates. The structural formula of a unit cell for a three-layer mineral $[Al_2(OH)_2(Si_2O_5)_2]_2$ and by omitting the 12 oxygen atoms of the tetrahedral sheet, the unit of the octa-hedral cell becomes $Al_4Si_8O_8(OH)_4$ or per aluminum atom it would be $AlSi_2O_2OH$. If the Al to Si ratio is enriched to 3 the corresponding restricted octahedral unit formula would be $Al_4Si_{12}O_{12}$ or $AlSi_3O_3$ per Al atom.

The formula of a unit cell of a two-layer mineral is $[Al_2(OH)_4-(Si_2O_5)]_2$. Omitting the six oxygen atoms from the tetrahedral sheet would give the formula per octahedral cell, $Al_4Si_4O_4(OH)_8$ or $AlSiO(OH)_2$ per Al atom. Since the Al atom was in all cases in six-coordination, bonding involved oxo (O^{2-}), hydroxo (OH^-), and aquo (H_2O) ligands.

In the case of the amorphous aluminum silicate preparations, negative charge was suggested to have developed from oxo ligands of the Al–O–Si and Al–OH–O–Si structures as illustrated below.

(1) Predominantly CECc, $AlSi_3O_3$ per Al atom.

$$
\begin{array}{ccccc}
\backslash|/ & \backslash|/ & \backslash|/ & & \backslash|/ \\
Si & Si & Si & & Si \\
| & | & | & & | \\
O & O & O^{.5-} & & O \\
\diagdown\!Al\diagup & Al & H^+ & + KCl \rightleftharpoons \ Al & K^+ \quad + HCl \\
| & | & | & & | \\
O & O & O^{.5-} & & O \\
| & | & | & & | \\
Si & Si & Si & & Si
\end{array}
$$

(2) Predominantly CECv, $AlSiO(OH)_2$ per Al atom.

$$
\left[
\begin{array}{cccccc}
\backslash|/ & & \backslash|/ & & & \backslash|/ \\
Si & & Si & & & Si \\
| & & | & & & | \\
O & OH & O^{.5-} & & & O \\
Al & Al & H & + KOH \rightleftharpoons & Al & K^+ \quad + H_2O \\
| & | & | & & | & \\
OH & OH & OH & & OH &
\end{array}
\right]_2
$$

In reaction (1) the Al:Si ratio was taken as 1:3 rather than the conventional 1:2 (since CECc was highest when the ratio was 1:3) and all of the coordination bonds involve oxo (O^{2-}) ligands. Two of the shared oxo ligands develop 0.5^- charge which in an acid medium share one-half bond each with H^+. This form of acidity, including $[Al(H_2O)_6]^{3+}$, is expected to be exchangeable by cations of neutral salts.

Reaction (2) depicts an Al:Si ratio of 1:1 and two hydroxo (OH^-) ligands combining in acid medium to form H and OH sharing Al–O–Si and Al–OH$_2$ bonds. Under the dominance of the OH ligands hydrogen is expected to be weakly dissociated and develops negative charge only on addition of base as shown. Exchangeable Al^{3+} in reaction (1) would be held similar to clay mineral bonding (Schwertman, 1963).

Histosols

The charge properties of histosols reported in Table 3 show approximately one-third of CECt as CECc. This is in agreement with a large number of results obtained by the author (unpublished). There is thus far little information based on the separation of CECc, CECv, and ASC. The latter is generally small except where the clay fraction is high.

Sorption of cations from neutral salts by H-humus was reported by Wiklander (1965) in the following order:

$$0.1N \; Li < Na < K < Mg < Ca < Sr < Ba < La$$
$$1.0N \; K < Na < Li < Mg < Ca = Sr = Ba < La$$

Coleman and Thomas (1967) summarized the general knowledge of the charge properties of organic matter. In early work, Broadbent and Bradford (1952) estimated that 55% of the CEC of organic matter from soil is accounted for by carboxyls. Phenolic and enolic groups contributed another 35%, and imide nitrogen, 10%. Schnitzer and Skinner (1963) also attributed 55% of the CEC of organic matter from a podzol to carboxyl groups. The remainder was due to phenolic and enolic groups. The observed CECc of the organic soils are, therefore, probably due to strongly dissociating carboxyl groups.

Soils High in Anion Sorption Capacity

Evidence indicates that anion sorption capacity (ASC) is high with mineral soils having high CECv and it is low with mineral soils high in CECc and with histosols high in organic matter. Although $[Al(OH_2)_6]^{3+}$ is a major component of CECc and this form of acidity is removed by leaching with 0.6N $BaCl_2$ it is not likely to be a factor in the determination of ASC with H_3PO_4 or $Ca(H_2PO_4)_2$. However, H_3PO_4, KH_2PO_4, K_2HPO_4, K_3PO_4, and several other anions were used to study their influence on CECc, CECv, and CECt with a Typic Rhodudult (Dyke), B33 horizon from the mountain region of North Carolina (Mehlich, 1964a, 1964b). The predominant mineral was kaolinite, followed by less than 10% each of vermiculite/hydrous mica and gibbsite. Charge properties of untreated soil were: CECc 0.9, CECv 7.1, CECt 8.0, and ASC 16.8 meq/100 g. Corresponding charge distribution was 11, 89, and 210% of CECt. Total sorbed SO_4^{2-} was 5.8 meq/100 g.

Ensuing charge properties following addition of 30 meq KOH or phosphate anion species and other anions to the soil are recorded in Table 4. At this rate CECc was somewhat but consistently higher with phosphate than with KOH. The results also show that CECv and CECt increased in the order: KOH < K_3PO_4 < K_2HPO_4 < KH_2PO_4, and H_3PO_4.

Influence of other anions on charge properties of the Typic Rhodudult soil are also recorded in Table 4. With 30 meq of anion added CECc was increased above the $Ca(OH)_2$ treatment by all compounds, except NH_4F and EDTA while CECv was increased mainly by $(HPO_3)_x$ and followed in decreasing order by NH_4VO_3, Na_2SiO_3, $(NH_4)_6Mo_7O_{24}$, and $NaBO_2$.

Table 4. Influence of anions on charge properties of soil high in sorbed sulfate and anion sorption capacity.

Anion added					Anion added				
30 meq/100 g	CECc	CECv	CECt	SO$_4$-S†	30 meq/100 g	CECc	CECv	CECt	SO$_4$-S†
	———— meq/100 g ————					———— meq/100 g ————			
Control	0.9	7.1	8.0	--	Ca(OH)	1.9	8.1	10.1	3.8
KOH	2.0	8.2	10.2	4.1	Na$_2$SiO$_3$	4.0	12.5	16.5	5.2
K$_3$PO$_4$	3.6	9.6	13.2	5.2	NH$_4$VO$_3$	3.8	14.3	18.1	5.2
K$_2$HPO$_4$	3.8	14.1	17.9	4.7	(NH$_4$)$_6$Mo$_7$O$_{24}$	3.8	11.0	14.8	4.3
KH$_2$PO$_4$	4.0	17.6	21.6	2.7	NaBO$_2$	3.0	9.3	12.3	2.7
H$_3$PO$_4$	4.2	17.6	21.8	0.7	NH$_4$F	2.0	7.1	9.1	3.8
(HPO$_3$)$_X$	4.4	17.9	22.3	2.7	EDTA	2.0	7.2	9.2	3.9

† Meq SO$_4$-S desorbed with addition of 30 meq anion.

The influence of various anions added at the rate of 30 meq/100 g on desorption of S is also shown in Table 4. Desorption is highest with K$_3$PO$_4$, Na$_2$SiO$_3$, NH$_4$VO$_3$ followed by K$_2$HPO$_4$, (NH$_4$)$_6$Mo$_7$O$_{24}$, KOH, EDTA, Ca(OH)$_2$, NH$_4$F, KH$_2$PO$_4$, (HPO$_3$)$_X$, NaBO$_2$, and H$_3$PO$_4$. Interpretation pretaining to these interactions are given in the following sections.

SORPTION OF PHOSPHATE AND DESORPTION OF SULFATE

Sulfate sorption by soils is known to increase with decreasing pH, while the reverse is obtained with respect to desorption (Ensminger, 1954; Harward et al., 1962; Aylmore et al., 1967; Chang and Thomas, 1963; Kamprath et al., 1956; Wild, 1940; Schoen, 1953; Rajan, 1979; Parfitt and Smart, 1978).

Sorption of phosphate by soils and minerals has been extensively investigated and the various factors influencing sorption have been studied by Barrow (1972), Bache (1963), Coleman et al. (1960), Dean and Rubins (1947), Fried and Dean (1955), Kittrick and Jackson (1956), Pugh (1934), Wild (1953), Wada (1959), White and Taylor (1977), Parfitt (1979), and others. The reactions involved are in general similar to those of sulfate (Harward and Reisenauer, 1966; Harward et al., 1962), however, the order of adsorption is considered highest for phosphate (Veith and Sposito, 1977) followed by molybdate, sulfate, and chloride = nitrate. With the exception of Gebhardt and Coleman (1974a, 1974b, 1974c) investigators in general did not determine sorption of anion in conjunction with sorption of hydronium or to measure residual acidity as a result of these treatments. Data presented in Fig. 1 to 4 provide some of this type of information involving phosphate.

Influence of Orthophosphoric Acid and its Salts

Studies were carried out with the B33 horizon of a Typic Rhodudult described in the previous section. Sorption of phosphate (Fig. 1) from H$_3$PO$_4$ was linear with concentrations up to 20 meq and subsequently in-

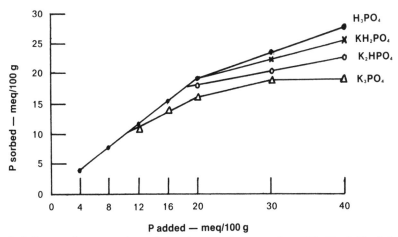

Fig. 1. Influence of concentration and phosphate species on sorption of P by Typic Rhodudult soil.

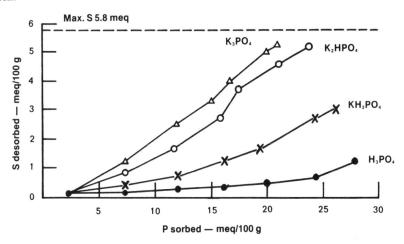

Fig. 2. Influence of concentration and phosphate species on desorption of sulfate from Typic Rhodudult soil.

creased curvilinearly from 20 meq up to 40 meq P added. The corresponding linear maximum was 16, 14, and 12 meq with KH_2PO_4, K_2HPO_4, and K_3PO_4, respectively, while the maximum at the 40 meq rate was 26, 23, and 20 meq.

The relationship between sorbed P and desorbed SO_4-S is shown in Fig. 2. Desorption of S by H_3PO_4 at all concentrations was small, but increased greatly in the order KH_2PO_4, K_2HPO_4, and K_3PO_4. With 20.5 meq of sorbed P from K_3PO_4, desorbed S was 5.2 or 90% of the 5.8 meq SO_4^{2-} present. Desorption of S with K_2HPO_4 paralleled desorption by K_3PO_4 but to a lesser extent. The predominantly acid forms of phosphate desorbed SO_4-S poorly, particularly at the lower concentrations of P with H_3PO_4. At the highest levels of sorbed P, desorption of S by H_3PO_4 was only 1.2 meq or 21% of the 5.8 meq S present in the soil.

Phosphorus-pH relationships are presented in Fig. 3. With the first increment of H_3PO_4 pH was lowered from 5.5 to 3.4 and continued to decrease to pH 2.9 with the highest rate. Addition of KH_2PO_4 lowered pH slightly to a constant of about 4.6, while with K_2HPO_4 pH initially decreased slightly, then increased sharply to pH 6.9 with 20 meq P added and then increased gradually to pH 7.3 up to maximum addition of P. The highest increases were with K_3PO_4 and reached a maximum of pH 8.9

In addition to pH, species of P added and SO_4-S desorbed were measured to pH 8.2 with $BaCl_2$-TEA for total acidity (ACt). These data are plotted in Fig. 4. Addition of H_3PO_4 increased ACt from 12.2 to 13.6 meq, and increased it further after 20 meq P were added up to 16 meq ACt. Additions of KH_2PO_4 decreased ACt initially and slightly with increasing sorption of P above 18.6 meq added. Phosphate sorbed from K_2HPO_4 decreased ACt sharply from 12.2 to 7.2 meq and then gradually to 4.8 meq at the highest rate of K_2HPO_4. With K_3PO_4 ACt decreased with increasing sorption of P to a maximum at 21 meq P, corresponding to 1.2 meq ACt.

The influence of phosphorus on charge properties, pH, ACt, sorption of P, and desorption of S is advantageously interpretative on the physicochemical properties of the orthophosphoric acid and its salts used (Nebergall et al., 1972). Orthophosphoric acid is tetrahedral and forms three series of salts corresponding to three stages of ionization as follows:

1) primary ionization; $H_3PO_4 \rightleftharpoons H^+ + H_2PO_4^-$
2) secondary ionization; $H_2PO_4^- \rightleftharpoons H^+ + HPO_4^{2-}$
3) tertiary ionization; $HPO_4^{2-} \rightleftharpoons H^+ + PO_3^{3-}$

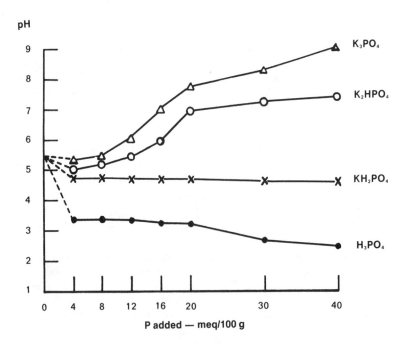

Fig. 3. Influence of concentration and phosphate species on pH of Typic Rhodudult soil.

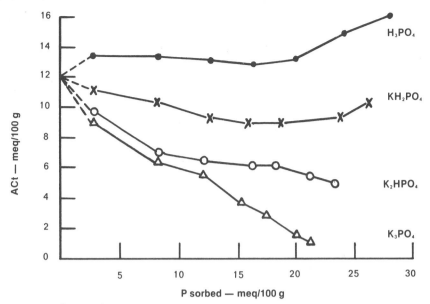

Fig. 4. Influence of concentration and phosphate species on neutralizable acidity (ACt) of Typic Rhodudult soil.

Influence of Hydroxyl on Desorption of Phosphate

Potentiometric titrations of aluminum and iron chlorides with sodium hydroxide in the presence of KH_2PO_4 were carried out by Swenson, Cole, and Sieling (1949). Results showed that for each metal ion two hydroxyls and one phosphate ion reacted to form $Al(H_2O)_3(OH)_2H_2PO_4$ or $Fe(H_2O)_3(OH)_2H_2PO_4$. The mechanism of phosphate sorption by aluminum was presented by the following equation:

$$Al(H_2O)_3 \begin{array}{c} OH \\ < OH \\ OH \end{array} + H_2PO_4^- \rightleftharpoons Al(H_2O)_3 \begin{array}{c} OH \\ OH - \\ H_2PO_4 \end{array} + OH^-$$

A general equilibrium reaction involving surface $H_2PO_4^-$ ion being held by clay minerals or hydrous oxides and OH ions was proposed in 1946 by Dean and Rubins as follows (S = surface):

$$S - H_2PO_4^- + OH^- \rightleftharpoons S - OH^- + H_2PO_4^-$$

It is noteworthy that one of the products in both illustrations was the $H_2PO_4^-$ ion.

Bache (1963) proposed the following equation for the hydrolytic decomposition of aluminum and iron phosphates:

$$AlPO_4 \cdot 2H_2O \rightleftharpoons Al(OH)_3 + H_3O^+ + H_2PO_4^-$$

$$FePO_4 \cdot 2H_2O \rightleftharpoons FeO \cdot OH + H_3O^+ + H_2PO_4^-$$

The equation by Swenson et al. and by Dean and Rubin depict an exchange reaction between OH^- and $H_2PO_4^-$, while those of Bache show $H_2PO_4^-$ in solution as a product of hydrolysis. Additional information regarding these reactions is provided by unpublished data of the author with $AlCl_3$ and $FeCl_3$ presented in Fig. 5 and 6.

In these experiments 1 mmol of $AlCl_3$ or $FeCl_3$ was titrated 1) with standard NaOH, 2) with 1 mmol of H_3PO_4, 3) with 1 mmol H_3PO_4 plus 2 mmol $BaCl_2$.

Aluminum or iron chloride titrated alone or titrated in the presence of 1 mmol H_3PO_4 or in the presence of $BaCl_2$ required 3 meq NaOH per mmol of either $AlCl_3$ or $FeCl_3$ for completion. Neutralization was above pH 6.0 and 5.0 for precipitation of $Al(H_2O)_3(OH)_3$ and $Fe(H_2O)_3(OH)_3$, respectively. In the presence of H_3PO_4 neutralization with base to form $Al(H_2O)_3(OH)_2H_2PO_4$ or the Fe analog was above pH 4.0 and 3.5, respectively. Addition of H_3PO_4 to $AlCl_3$ reduced pH from 3.5 to 1.8, and with $FeCl_3$ pH was reduced from 2.1 to 1.9.

The titration with $AlCl_3$ plus H_3PO_4 and $FeCl_3$ plus H_3PO_4 were repeated and terminated when 3.5, 4.0, and 4.5 meq NaOH were added. The suspensions were transferred to centrifuge tubes, warmed to facilitate aggregation, centrifuged, and P in solution determined by the vanadomolybdophosphate method. These techniques were repeated and the centrifugates from the 3.5, 4.0, and 4.5 ml NaOH increments titrated with HCl for the purpose of identifying the hydrolysis product of $Al(H_2O)_3(OH)_2H_2PO_4$ and $Fe(H_2O)_3(OH)_2H_2PO_4$. The extracts of the three NaOH increments are labeled 0.5, 1.0, and 1.5 meq NaOH indicating the excess of base above the 3 meq required for completion of the reaction. The results shown in Fig. 6 reveal typical titration characteristics of dibasic sodium phosphate with HCl. The quantity of P in solution as $NaHPO_4$ based on the inflection point at pH 4.0 was also compared with the total amount of P in solution.

The results in Table 5 show that with addition of 0.5 meq NaOH, 67% of the P in solution was HPO_4^{2-} while 33% was probably $H_2PO_4^-$. With 1.0 and 1.5 meq excess NaOH added 97 to 100% was HPO_4^{2-} in the Al series and 95 to 100% in the Fe series. Evidently, these observations do not support the quoted literature suggestions that for the exchange of one equivalent $H_2PO_4^-$ per Al or Fe atom only one OH^- is required. Present evidence suggests therefore the reaction:

$$Al(H_2O)_3(OH)_2H_2PO_4 + 2OH^- \rightleftharpoons Al(H_2O)_3(OH)_3 + HPO_4^{2-} + HOH$$

Table 5. Quantity of HPO_4^{2-} in solution following desorption of P by hydroxyl from aluminum and iron dihydroxydihydrogen phosphate.

NaOH added,	0.5	1.0	1.5	0.5	1.0	1.5
meq/mmole Al or Fe compound	$Al(H_2O)_3(OH)_2H_2PO_4$			$Fe(H_2O)_3(OH)_2H_2PO_4$		
Total P, t†	0.54	1.11	1.86	0.63	1.26	1.98
P as Na_2HPO_4, t	0.36	1.08	1.92	0.4	1.20	2.04
P as Na_2HPO_4, %	67	97	100	67	95	100

† t = meq P/mmole Al or Fe phosphate.

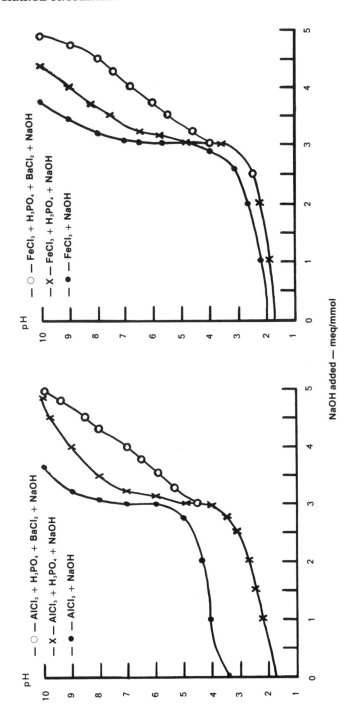

Fig. 5. Base-pH titration characteristics of Al or Fe salts with and without additions of H_3PO_4 and $BaCl_2$.

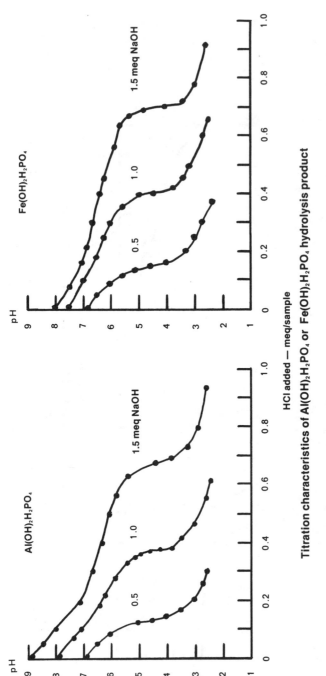

Fig. 6. Titration characteristics of Al(H₂O)₃(OH)₂H₂PO₄ or Fe(H₂O)₃(OH)₂H₂PO₄ products with additions of 0.5, 1.0, and 1.5 meq excess NaOH.

A similar equation would apply to the iron dihydroxydihydrogen phosphate.

In the presence of $BaCl_2$ the supernatant centrifugate contained no detectable P above the inflection point where more than 3 meq hydroxyl were added (Fig. 5). The titration curves have greater buffer tendency due to formation of water insoluble $BaHPO_4$ and possibly $Ba_3(PO_4)_2$. The same effects would be expected to occur with $CaCl_2$.

Tri, di, monobasic phosphates, and ortho phosphoric acid have been shown to influence charge properties, neutralization of soil acidity, and desorption of sulfate. Evidently, formation of sesquidihydroxide dihydrogen phosphate proceeds rapidly when the reactants have equivalent positive and negative charges. Hence, $[Al(H_2O)_6]^{3+}$ reacts quantitatively with PO_4^{3-}, $[Al(H_2O)_5(OH)]^{2+}$ with HPO_4^{2-}, and $[Al(H_2O)_4(OH)_2]^+$ with $H_2PO_4^-$ and finally,

$$Al(H_2O)_3(OH)_3 + H_3PO_4 = Al(H_2O)_3(OH)_2H_2PO_4 + H_2O$$

The same reaction would occur with Fe^{3+} species.

Mechanism of Positive Charge and Anion Sorption Capacity

Positive charge in gibbsite was attributed by Hingston et al. (1967) to non-specific anion adsorption when $Al(OH)_3$ was protonated with HCl to yield $[Al(OH)_2(H_2O—Cl)]^+$. Non-specific exchange occurred with NO_3^- to form $[Al(OH)_2(H_2O—NO_3^-)] + Cl^-$. Neutralization of the positive charge with addition of $H_2PO_4^-$ gave rise to anions being specifically adsorbed at a given pH since they could not be desorbed by non-specifically adsorbed anions at the same concentration and pH (Muljadi et al., 1966). There is general agreement that protonation of $Al(H_2O)_3(OH)_3$ or $Fe(H_2O)_3(OH)_3$ involves one hydroxy group.

If the acid involved was monoprotic (HCl, HNO_3), diprotic (H_2SO_4), or triprotic (H_3PO_4) and since both H_2SO_4 and H_3PO_4 have tetrahedral structure, one hydroxo ligand would be expected to exchange for one oxo ligand rather than an HSO_4^- or $H_2PO_4^-$ anion. Furthermore, if the anion is Cl^- or NO_3^-, one OH^- would be needed to form $Al(H_2O)_3(OH)_3$, or $Fe(H_2O)_3(OH)_3$, while 2 OH^- and 3 OH^- would be required if the sorbed anions were HSO_4^- or $H_2PO_4^-$, respectively. Consequently, positive charge would develop when reaction was with monoprotic acid while anion sorption capacity would develop when the reaction involved diprotic or triprotic acids. On an equivalent basis, the hydroxyl requirements would be 0.33, 0.66, and 1.0 meq per equivalent of trivalent aluminum or iron. When expressed on an equivalent rather than a molar basis, CECv and ASC of gibbsite and goethite in meq/100 g were of the same magnitude (Table 2), while ASC when expressed on a molar basis is one-third of ASC.

Uniformity and clarity are expected to reign in the field of surface charge of soils when all data are expressed on a meq rather than cations on an equivalent and anions on a mixed meq or mmol basis.

INFLUENCE OF VARIOUS ANIONS ON DESORPTION OF
SULFATE, pH, ACIDITY, AND CHARGE
PROPERTIES OF SOILS

The Typic Rhodudult soil was also used to study the effect of various other anions on desorption of sulfate, pH, acidity, and charge properties. Results in Fig. 7 showed desorption of S increased with increasing concentration and in the following order: $CH_3COONa < NaBO_2 < KOH = EDTA < Na_2SiO_3$, and desorption data in Fig. 8 increased in the order: metaphosphoric acid $(HPO_3)_X < NH_4F < (NH_4)_6Mo_7O_{24} < NH_4VO_3$.

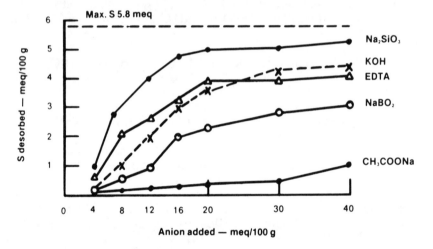

Fig. 7. Desorption of sulfate from Typic Rhodudult soil by EDTA, salts and bases.

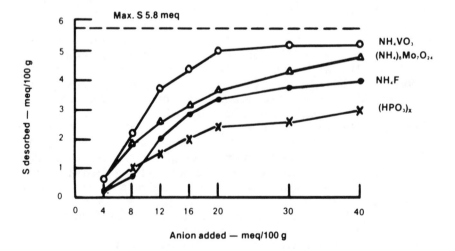

Fig. 8. Desorption of sulfate from Typic Rhodudult soil by various compounds.

Sodium acetate, within the concentration used has low capacity for the desorption of sulfate and behaved essentially as a neutral salt. Increased desorption of sulfate occurred with anions which increased pH (Fig. 9) or decreased $BaCl_2$-TEA titratable acidity (ACt) in Fig. 10. These include KOH, Na_2SiO_3, $NaBO_2$, and NH_4F. Within this group, highest desorption of sulfate was with Na_2SiO_3, followed by KOH, NH_4F, and $NaBO_2$. Judging from the charge properties in Table 4, CECc and CECv were higher with Na_2SiO_3 than with $NaBO_2$ and particularly KOH and NH_4F. The high desorption of sulfate and increased negative charge is apparently due to the higher valency of the sulfate anion and entry of an oxo ligand into the coordination sphere of the Al and Fe atoms. Ammonium fluoride initially decreased ACt and above addition of 12 meq ACt gradually increased. This increase coincided with a leveling off of desorption of S.

According to Hingston et al., 1973; and Parfitt, 1979, the fluoride ion reacts with goethite by ligand exchange while with gibbsite a large number of OH groups are replaced, suggesting that Al•OH•Al as well as Al•OH groups can exchange with fluoride. The fluoride ion also reacts with $Al(OH)_3$ and Al-phosphates to form the $(NH_4)_3AlF_6$ complex (Swenson et al., 1949; Turner and Rice, 1952; Jackson, 1963). A corresponding trivalent fluoroferrate ion was formed in acid solution but was destroyed in neutral to alkaline solution where $Fe(OH)_3$ was the stable form.

Calcium and potassium hydroxide desorb sulfate from soil to form $CaSO_4$ and K_2SO_4, respectively. Formation of $CaSO_4$ was shown by Mehlich (1964a, 1964b) to require a higher rate of $Ca(OH)_2$ as compared to soil low in sorbed sulfate. The excess OH ions from $Ca(OH)_2$ consumed in the desorption of S was also reflected in an increase in residual acidity.

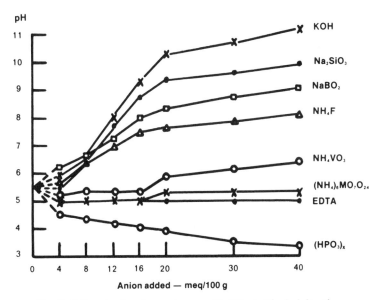

Fig. 9. Influence of various anions on pH of Typic Rhodudult soil.

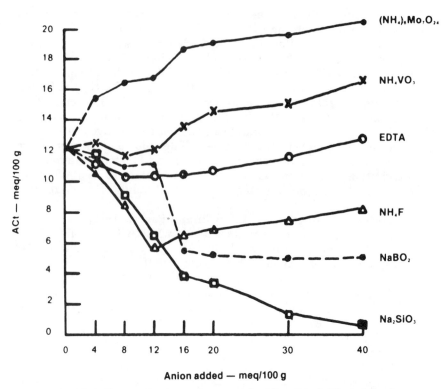

Fig. 10. Influence of various anions on total acidity of Typic Rhodudult soil.

Desorption of S by ethylenediaminetetraaetic acid (EDTA) was within the same magnitude as KOH (Fig. 7), although the pH was little influenced by concentration (Fig. 9) while ACt decreased somewhat at the lower rates (Fig. 10). The effect of EDTA on charge properties (Table 4) is small and about the same as KOH.

The results in Fig. 8 showed high desorption of sulfate by ammonium vanadate, followed by ammonium molybdate and metaphosphoric acid $(HPO_3)_x$. Vanadate and molybdate had little effect on pH while the pH with metaphosphate decreased with increasing concentration (Fig. 9). The latter had little influence on ACt (data not presented) while NH_4VO_3 and particularly $(NH_4)_6Mo_7O_{24}$ substantially increased ACt with increasing concentration (Fig. 10). Judging from data in Table 4, all three anions increased CECc above the KOH effect, while CECv increased in the order molybdate, vanadate, and metaphosphate. The influence of the latter was similar to those of KH_2PO_4 and H_3PO_4 and NH_4VO_3 related more to K_2HPO_4.

The high efficiency of NH_4VO_3 in the desorption of S cannot be explained on ligand exchange through hydroxyl since it was accompanied by protonation (increasing ACt). The same reaction was observed with molybdate. A tentative suggestion for this apparent anomaly could be due to possible chelation characteristics. The data show, however, that

NH_4VO_3 has great efficiency in the desorption of sulfate and is highly comparable with those of K_3PO_4, K_2HPO_4, and Na_2SiO_3.

SUMMARY AND CONCLUSION

Knowledge of charge properties of soils has great importance in soil classification, soil fertility, and especially with respect to sorption and desorption of cations and anions. The charge properties involve 1) constant negative charge cation exchange capacity (CECc), 2) variable negative charge cation exchange capacity (CECv), 3) total negative charge cation exchange capacity (CECt), 4) positive charge anion exchange capacity (AEC), and 5) anion sorption capacity (ASC). In the proposed procedure CECc is the quantity of $BaCl_2$ exchangeable H^+/Al^{3+} of the H/Al saturated soil, CECt is the quantity of exchangeable Ba which was absorbed following saturation of soil with $BaCl_2$ followed by $BaCl_2$-TEA (triethanolamine) at pH 8.2, while CECv is the calculated difference between CECt-CECc. The consumption of base by the soil, which was supplied by $BaCl_2$-TEA following treatment with $BaCl_2$, represents neutralization of variable negative charge acidity. This acidity is equivalent to CECv. However, if the soil also has AEC or ASC, the base consumed includes sorption of hydroxyl in exchange for anions. Consequently, the difference between residual acidity and CECv is for routine purposes equivalent to ASC. Positive charge is determined on a separate sample where the quantity of Cl^- exchanged by another anion (NO_3^-) from the H^+/Al^{3+} saturated soil is equal to AEC. All negative and positive charge data including sorption and desorption of anions are expressed on an equivalent basis.

Sorption of cations are substantially influenced by type of mineral and organic colloids. Divalent cations are generally more strongly sorbed than monovalent cations. Sorption of cations from unbuffered salts is high by three-layer silicate minerals, followed by two layer silicate minerals, soils high in sesquioxidehydrates, allophane, and organic matter. The converse order occurs with respect to cations associated with CECv.

The origin of sorbed anions is primarily the aluminum and iron hydroxides or their coatings on clay surfaces. Sorption of monovalent anions results through exchange of hydroxo ligands for aquo ligands by monoprotic acid (HCl, HNO_3). The aquo ligands countered by monovalent anions are quantitatively exchanged by salts of other anions. This exchange accounts primarily for positive charge (AEC). Dehydroxylation with diprotic acid (H_2SO_4), or desorption of aquo ligands leads to entry of oxo ligands, countered by the HSO_4^- anion. Desorption of HSO_4^- anion proceeds by addition of hydroxyl or chelating chemicals, such as EDTA. Acids, including H_3PO_4, or unbuffered salts are ineffective and desorption of HSO_4^- increases in the order $H_2PO_4^-$, HPO_4^{2-}, and PO_4^{3-}, while Na_2SiO_3 and NH_4VO_3 are also highly effective. Addition of triprotic acid (H_3PO_4) or desorption of aquo ligands leads to entry of oxo ligands countered by $H_2PO_4^-$ anion. Sorption of P is highest with H_3PO_4 and decreases with $H_2PO_4^-$, HPO_4^{2-}, and PO_4^{3-}. Desorption of $H_2PO_4^-$ from $Al(H_2O)_3(OH)_2H_2PO_4$ or $Fe(H_2O)_3(OH)_2H_2PO_4$ proceeds with addition of hydroxyl by formation of $HPO_4{}^{2-}$, while on hydrolysis the ionization products are $H_2PO_4{}^-$ and H_3O^+.

There has been great progress relating to the mechanism of sorption and desorption of cations and anions, while the chemical characterization of charge properties of soils require greater practical application, particularly those concerning reactions with waste products.

REFERENCES

1. Aomine, S., and M. L. Jackson. 1959. Allophane determination in Ando soils by cation-exchange delta value. Soil Sci. Soc. Amer. Proc. 23:210–214.

2. Aylmore, L. A. G., M. Karin, and J. P. Quirk. 1967. Adsorption and desorption of sulfate ions by soil constituents. Soil Sci. 103:10–15.

3. Bache, B. W. 1963. Aluminum and iron phosphate studies relating to soils. I. Solution and hydrolysis of variscite and strengite. J. Soil Sci. 14:113–123.

4. Barrow, N. J. 1972. Influence of solution concentration of calcium on the adsorption of phosphate, sulfate, and molybdate by soils. Soil Sci. 113:175–180.

5. Bhumbla, D. R., and E. O. McLean. 1965. Aluminum in soils. VI. Changes in pH-dependent acidity, cation-exchange capacity, and extractable aluminum with additions of lime to acid surface soils. Soil Sci. Soc. Amer. Proc. 29:370–374.

6. Birrell, K. S. 1958. Reactions of amorphous soil colloids with ions in solution. N. Z. Soc. Soil Sci. Proc. 3:11–13.

7. ————, and M. Gradwell. 1956. Ion-exchange phenomena in some soils containing amorphous mineral constituents. J. Soil Sci. 7:130–147.

8. Broadbent, F. E., and G. R. Bradford. 1952. Cation-exchange groupings in the soil organic fraction. Soil Sci. 74:447–457.

9. Chang, M. L., and G. W. Thomas. 1963. A suggested mechanism for sulfate adsorption by soils. Soil Sci. Soc. Amer. Proc. 27:281–283.

10. Chatterjee. B. 1956. The variation of cationic activity in colloidal clays with concentration of the disperse phase. J. Indian Chem. Co. 33:399.

11. ————, and C. E. Marshall. 1950. Studies in the ionization of magnesium, calcium, and barium clays. J. Phys. Coll. Chem. 54:671.

12. Coleman, N. T., and G. W. Thomas. 1964. Buffer curves of acid clays as affected by the presence of ferric iron and aluminum. Soil Sci. Soc. Amer. Proc. 28:187–190.

13. ————, S. B. Weed, and R. J. McCracken. 1959. Cation-exchange capacity and exchangeable cations in Piedmont soils of North Carolina. Soil Sci. Soc. Amer. Proc. 23: 146–149.

14. ————, and G. W. Thomas. 1967. The basic chemistry of soil acidity. In R. W. Pearson and F. Adams (ed.) Soil acidity and liming. Agronomy 12:1–41. Am. Soc. of Agron., Madison, Wis.

15. ————, J. T. Thorup, and W. A. Jackson. 1960. Phosphate sorption reactions that involve exchangeable aluminum. Soil Soc. 90:1–7.

16. Dean, L. A., and E. J. Rubins. 1947. Anion exchange in soils. I. Exchangeable phosphorus and the anion exchange capacity. Soil Sci. 63:377–387.

17. deVilliers, J. M., and M. L. Jackson. 1967. Cation exchange capacity variations with pH in soil clays. Soil Sci. Soc. Amer. Proc. 31:173–176.

18. Dolman, J. D., and S. W. Buol. 1967. A study of organic soils (Histosols). Tech. Bul. No. 181. North Carolina Agric. Exp. Stn.

19. Ensminger, L. E. 1954. Some factors affecting the adsorption of sulfate by Alabama soils. Soil Sci. Soc. Am. Proc. 18:259–264.

20. Fieldes, M. 1955. Mechanisms of ion adsorption by inorganic soil colloids. N. Z. J. Sci. 3:563–579.

21. ————, and R. K. Schofield. 1960. Mechanism of ion adsorption by inorganic soil colloids. N. Z. J. Sci. 3:563–579.

22. Fried, M., and L. A. Dean. 1955. Phosphate retention by iron and aluminum in cation exchange systems. Soil Sci. Soc. Am. Proc. 19:143–147.

23. Gebhardt, H., and N. T. Coleman. 1974a. Anion adsorption by allophanic tropical soils. I. Chloride adsorption. Soil Sci. Soc. Am. Proc. 38:255–259.

24. ————, and ————. 1974b. Anion adsorption by allophanic tropical soils. II. Sulfate adsorption. Soil Sci. Soc. Am. Proc. 38:259–262.

25. ————, and ————. 1974c. Anion adsorption by allophanic tropical soils. III. Phosphate adsorption. Soil Sci. Soc. Am. Proc. 38:263–266.

26. Gieseking, J. E. 1939. Mechanism of cation exchange in the montmorillonite-beidel-lite-nontronite type of clay minerals. Soil Sci. 47:1–13.

27. ————. 1949. The clay minerals in soils. Adv. Agron. 1:159–204.

28. ————, and H. Jenny. 1936. Behavior of polyvalent cations in base exchange. Soil Sci. 42:273–280.

29. Gillam, W. S. 1940. A study on the chemical nature of humic acid. Soil Sci. 49:433–453.

30. Goates, J. R., and K. Anderson. 1956. Acidic properties of quartz. Soil Sci. 81:277–282.

31. Harward, M. E., and H. M. Reisenauer. 1966. Reactions and movement of inorganic soil sulfur. Soil Sci. 101:326–335.

32. ————, T. T. Chao, and S. C. Fang. 1962. Soil properties and constituents in relation to mechanisms of sulfate adsorption. p. 93–113. Radioisotopes in soil-plant nutrition studies. Proc. Int. Atomic Energy Agency, Vienna.

33. Helling, C. S., G. Chesters, and R. B. Corey. 1964. Contributions of organic matter and clay to soil cation-exchange capacity as affected by the pH of the saturating solution. Soil Sci. Soc. Amer. Proc. 28:517–520.

34. Hingston, F. J., R. J. Atkinson, A. M. Posner, and J. R. Quirk. 1967. Specific adsorption of anions. Nature 215:1459–1461.

35. ————, A. M. Posner, and J. P. Quirk. 1972. Anion adsorption by goethite and gibbsite. I. The role of the proton in determining adsorption envelopes. J. Soil Sci. 23:177–193.

36. Hofmann, U., K. Endell, and D. Wilm. 1933. Kristallstruktur and Quellung von Montmorillonit. Z. Krist. 86:340–348.

37. Jackson, M. L. 1963. Aluminum bonding in soils: A unifying principle in soil science. Soil Sci. Soc. Amer. Proc. 27:1–10.

38. Jarusov, S. S. 1937. Mobility of exchangeable cations in the soil. Soil Sci. 43:285–303.

39. Jenny, H. 1932. Studies on the mechanism of ionic exchange in colloidal aluminum silicates. Phys. Chem. 36:2217–2258.

40. ————. 1961. Reflections on the soil acidity merry-go-round. Soil Sci. Soc. Amer. Proc. 25:428–432.

41. Kittrick, J. A., and M. L. Jackson. 1956. Electron-microscope observation of the reaction of phosphate with minerals, leading to a unified theory of phosphate fixation in soils. Soil Sci. 7:81–89.

42. Kamprath, E. J., W. L. Nelson, and J. W. Fitts. 1956. The effect of pH, sulfate, and phosphate concentrations on the adsorption of sulfate by soils. Soil Sci. Soc. Amer. Proc. 20:463–466.

43. Lewis, T. E., and F. E. Broadbent. 1961. Soil organic matter-metal complexes. 4. Nature and properties of exchange sites. Soil Sci. 91:393–399.

44. Low, P. F. 1955. The role of aluminum in the titration of bentonite. Soil Sci. Soc. Amer. Proc. 19:135–139.

45. McLean, E. O., D. C. Reicosky, and C. Lakshmanan. 1965. Aluminum in soils. VII. Interrelationships of organic matter, liming, and extractable aluminum with "permanent charge" (KCl) and pH-dependent cation-exchange capacity of surface soils. Soil Sci. Soc. Amer. Proc. 29:374–378.

46. Marshall, C. E. 1964. The physical chemistry and mineralogy of soils. Vol. 1. John Wiley and Sons, New York. 388 p.

47. Mattson, S. 1928. The electrokinette and chemical behaviour of the aluminosilicates. Soil Sci. 25:289–311.

48. ————. 1931. The laws of soil colloidal behaviour. V. Ion adsorption and exchange. Soil Sci. 31:311–331.

49. Mehlich, A. 1953. Rapid determination of cation and anion exchange properties and pHe of soils. J. Assoc. Offic. Agr. Chem. 36:447–457.

50. ————. 1960. Charge characterization of soils. 7th Int. Congr. Soil Sci. II. 292–302.

51. ————. 1964a. Influence of sorbed hydroxyl and sulfate on neutralization of soil acidity. Soil Sci. Soc. Amer. Proc. 28:492–496.

52. ————. 1964b. Influence of sorbed hydroxyl and sulfate on liming efficiency, pH, and conductivity. Soil Sci. Soc. Am. Proc. 28:496–499.

53. ————. 1967. Negative Ladungseigenschaften synthetischer Aluminum-silikate. Pflanzenernähr. Bodenkund. 117:193–204.

54. Muljadi, D., A. M. Posner, and J. P. Quirk. 1966. The mechanism of phosphate adsorption by kaolinite, gibbsite, and pseudoboehmite, I-III. J. Soil Sci. 17:212–247.

55. Nebergall, W. H., F. C. Schmidt, and H. F. Holtzclaw. 1972. General chemistry. 4th ed. D. C. Heath and Co., Lexington, Mass.

56. Parfitt, R. L. 1979. Anion adsorption by soils and soil materials. Adv. Agron. 30:1–50.

57. ————, and St. C. Smart. 1978. The mechanism of sulfate adsorption on iron oxides. Soil Sci. Soc. Am. J. 42:48–50.

58. Paver, H., and C. E. Marshall. 1934. The role of aluminum in the reactions of the clays. Chem. Ind. 740–760.

59. Pratt, P. F., and F. L. Bair. 1962. Cation-exchange properties of some acid soils of California. Hilgardia 33:689–706.

60. Pugh, A. J. 1934. Laws of soil colloidal behaviour. III. Colloidal phosphates. Soil Sci. 38:315–334.

61. Quirk, J. P. 1960. Negative and positive adsorption of chloride by kaolinite. Nature 188–253.

62. Rajan, S. S. S. 1979. Adsorption and desorption of sulfate and charge relationships in allophanic clays. Soil Sci. Soc. Am. J. 43:65–69.

63. Rich, C. I., and S. S. Obenshain. 1955. Chemical and clay mineral properties of a Red-Yellow Podzolic soil derived from mica schist. Soil Sci. Soc. Amer. Proc. 19:334–339.

64. Ross, C. J., K. Lawton, and B. G. Ellis. 1964. Lime requirement related to physical and chemical properties of nine Michigan soils. Soil Sci. Soc. Amer. Proc. 28:209–212.

65. Schachtschabel, P. 1940. In Scheffer-Schachtschabel. Lehrbuch der Bodenkunde; Ferdinand Enke Verlag, Stuttgart, 1966.

66. Schnitzer, M., and J. G. Desjardins. 1962. Molecular and equivalent weights of the organic matter of a Podzol. Soil Sci. Soc. Amer. Proc. 26:362–365.

67. ————, and U. C. Gupta. 1964. Chemical characteristics of the organic matter extracted from the O and B2 horizons of a Gray Wooded soil. Soil Sci. Soc. Amer. Proc. 28:374–377.

68. ————, and S. I. M. Skinner. 1963. Organo-metallic interactions in soils. 1. Reactions between a number of metal ions and the organic matter of a Podzol Bh horizon. Soil Sci. 96:86–93.

69. ————, and J. R. Wright. 1960. Nitric acid oxidation of the organic matter of a Podzol. Soil Sci. Soc. Amer. Proc. 24:273–277.

70. Schoen, U. 1953. Die Bindung von Anionen, besonders von phosphationen an Tonen —Eine Obersicht. Z. Planzenernähr. Düng. Bodenkd. 60(105):31–54.

71. Schofield, R. K. 1939. The electrical charges on clay particles. Soils Fert. 2:1–5.

72. ————. 1949. Effect of pH on electric charges carried by clay particles. J. Soil Sci. 1: 1–8.

73. Schwertmann, U. 1963. Das Verhalten des Austauschbaren Aluminums bei der Alterung von H-Tonmineralen. Mitteilungen der Deutschen Bodenkundlichen Gesellschaft Bd. 1: 319–323.

74. ————, and M. L. Jackson. 1964. Influence of hydroxy-aluminum ions on pH titration curves of hydronium-aluminum clays. Soil Sci. Soc. Amer. Proc. 28:179–182.

75. Sumner, M. E., and N. G. Reeve. 1966. The effect of iron oxide impurities on the positive and negative adsorption of chloride by kaolinites. J. Soil Sci. 17:274–279.

76. Swenson, R. M., C. V. Cole, and D. H. Sieling. 1949. Fixation of phosphorus by iron and aluminum and replacement by organic and inorganic ions. Soil Sci. 67:3–22.

77. Theisen, A. A. 1966a. Kristalline Bestandteile saurer tropischen Boden auf vulkanischem Ausgangs-material in Kenia. Z. Pflanzenernähr. Düng. Bodenkd. 115:173–181.

78. ————. 1966b. Rontgenamorphe Bestandteile saurer tropischer Boden auf vulkanischem Ausgangsmaterial in Kenia. Z. Pflanzenernähr. Düng. Bodenkd. 115:181–192.

79. Thomas, G. W. 1960. Forms of aluminum in cation exchangers. Int. Congr. Soil Sci., Trans. 7th, Madison, Wis. 2:364–369.

80. ————. 1977. Historical developments in soil chemistry: Ion exchange. Soil Sci. Soc. Amer. J. 41:230–238.

81. Turner, R. G., and H. M. Rice. 1952. Role of the fluoride ion in release of phosphate adsorbed by aluminum and iron hydroxides. Soil Sci. 74:141–148.

82. Vanselow, A. P. 1932. Equilibria of the base-exchange reaction of bentonites, permutites, soil colloids, and zeolites. Soil Sci. 33:98–113.

83. Veith, J. A., and G. Sposito. 1977. Reactions of aluminosilicates, aluminum hydrous oxides, and aluminum oxide with ophosphate; the formation of x-ray amorphous analogs of variscite and montebrasite. Soil Sci. Soc. Amer. J. 41:870–876.

84. Volk, V. V., and M. L. Jackson. 1964. Inorganic pH dependent cation exchange charge of soils. p. 281–285. In W. F. Bradley (ed.) Clays and clay minerals, Proc. 12th Conf. Pergamon Press, Ltd., New York.

85. Wada, K. 1959. Reactions of phosphate with allophane and halloysite. Soil Sci. 87:325–330.

86. ————, and H. Ataka. 1958. The ion uptake mechanism of allophane. Soil Plant Food (Tokyo) 4:12–18.

87. White, R. E., and A. W. Taylor. 1977. Effect of pH on phosphate adsorption and isotopic exchange in acid soils at low and high additions of soluble phosphate. J. Soil Sci. 28:48–61.

88. Wild, A. 1940. The retention of phosphate by soil. A review. J. Soil Sci. 1:221–238.

89. ————. 1953. The effect of exchangeable cations on the retention of phosphate by clay. J. Soil Sci. 4:72–85.

90. Wiklander, L. 1965. Cation and anion exchange phenomena. F. E. Bear (ed.) Chemistry of the soil, Ch. 4. Reinhold Publish. Corp., New York.

CHAPTER 5

Oxidation-Reduction Status of Aerobic Soils[1]

RICHMOND J. BARTLETT[2]

---

[1] Vermont Agric. Exp. Stn. Journal article no. 444.

[2] Univ. of Vermont, Burlington.

THE NATURAL SOIL SYSTEM

Soils in situ are living heterogeneous sytems, open to inputs and out-puts of heat, air, solutions, and solids. Soils researchers often are guilty of aggressively drying, sieving, mixing, dry-storing, and otherwise depredat-ing soils in order to convert them into homogenized standardized labora-tory reagents. While such material still retains many soil properties, the changes during drying and storage are extensive enough that perhaps now it might accurately be termed lab dirt.

Unfortunately, or fortunately, depending on one's point of view, we are not small enough to move about freely inside a 10-micron soil pore. Therefore, we must devise schemes that will enable us to mentally probe the darkness among the mineral surfaces and dead and living organic forms of varying viscosities and volatilities. Then we must figure out ways to probe smaller dimensions to become aware of the chemical happenings as oxidative and reductive soil environments come together at interfaces in peds, paddies, sediments, humic and clay surfaces, and living cell/soil boundaries.

Soils are balanced in a remarkably stable state of nonequilibrium. The physical chemist would label most soil ingredients meta-stable, which in plain language means they really are stable, for a while at least, but they are not supposed to be—if thermodynamics holds sway. Fortun-ately for life on earth, there are long periods of time during which thermodynamic equilibrium is not reached in aggregations we know as plants, animals, and soils. Thermodynamics tells us that existence of soils is precarious—like that of people. But experience tells us that soils and people are wonderfully resilient and resistant to disintegration.

Total equilibrium between soil and air would literally destroy the roots of life as all of the organic C reduced by the power of sunlight would be returned to stable CO_2. But if soil did not move continuously toward equilibrium with air at an adequate pace, the oxidized C in the atmos-phere would soon vanish. Factors fostering nonequilibrium are limited aeration porosity of soils and meta-stability of soil humus.

This paper deals with the behavior of aerobic soils, those usually con-taining free oxygen. The status of Fe serves as a convenient boundary for separating oxidizing from reducing conditions in soils (8). Oxidized Fe indicates an aerobic soil; reduced Fe, an anaerobic one, or an anaerobic zone (22, 57) within an aerobic soil. If the large pores of a soil contain a sustained level of O_2, and the temperature is above freezing, the soil will be either aerobic or in the process of becoming aerobic. Reduced species present will be changing into oxidized forms. In the process of gleization, Fe becomes reduced and soluble under saturated conditions inside a soil aggregate (60). On diffusing outward, the Fe becomes oxidized by O_2. The result is a reddish ped exterior and a gray ped interior.

SOIL ELECTRONS AND PROTONS

pe and pH

Oxidation is the donation of an electron; reduction is electron reception. As substances are oxidized, they become deficient in electrons. Reduced substances are electron rich. Availability of electrons is directly related to tendency for oxidation. Therefore, electron rich substances display strong inclinations toward oxidation. Electron poor substances lean toward reduction.

Following are the two most basic soil redox half reactions:

$$H^+ + \tfrac{1}{4} O_2 + e^- = \tfrac{1}{2} H_2O \qquad \text{reduction of } O_2 \qquad [1]$$

$$1/24 \text{ glucose} + \tfrac{1}{4} H_2O = \tfrac{1}{4} CO_2 + H^+ + e^- \text{ oxidation of C} \qquad [2]$$

Oxidation of H_2O accompanies reduction of CO_2 only in the process of photosynthesis. Neither of the above reactions as written is reversible in soils, except in cells containing chlorophyll. Consequently, fluctuations of H^+ and e^- activities in soils affect shifts in equilibria in these basic half-reactions in one direction only, from left to right in both.

Redox reactions always involve two half-reactions balanced so that the electron donated by the oxidation half is accepted by the reduction half leaving no free electrons. Most redox half-reactions in soils are reversible such as those following:

$$\tfrac{1}{2} H_2O_2 + H^+ + e^- = H_2O \qquad \text{reduction} \qquad [3]$$

$$\tfrac{1}{2} Mn^{2+} + H_2O = \tfrac{1}{2} MnO_2 + 2H^+ + e^- \quad \text{oxidation} \qquad [4]$$

Since in the coupling of Eq. 3 and 4 to give a redox reaction, there is a net release of H^+, we might expect the oxidation of Mn^{2+} by H_2O_2 to produce a lowering of pH and to be favored by high pH.

The following generalization of Eq. 3 shows the equilibrium constant K to be the same as the sum of pe and pH when reduced and oxidized species are equal and only 1 e^- and 1 H^+ per mole are involved:

$$K = \frac{(\text{red})}{(\text{ox}) \, (e^-) \, (H^+)} \qquad [5]$$

$$\log K - \log \frac{(\text{red})}{(\text{ox})} = - \log (e^-) - \log (H^+)$$

$$= \text{pe} + \text{pH} \qquad [6]$$
$$[\text{by definition of pe and pH (58)}]$$

Analogous to pH as an energy level of protons, indicating proton supplying intensity, pe expresses an energy level of electrons and indicates electron supplying intensity. A low pe system will tend to donate electrons, a high pe system, to receive them. A low pe tends to rise; a high pe tends to become lower. The farther apart the electrons, the more work required to bring them together and the higher the pe. The more concentrated the electrons, the lower the energy level and the pe. A unit of pe is equivalent in free energy to a pH unit, each representing 2.3RT calories per mole. (log K may be calculated by dividing the sums of the Gibbs free energies of the reactants in a half-reaction minus those of the products by 2.3RT.)

By representing the ratios of activities of all of the reduced and oxidized species in a redox system in relation to the sums of all of the equilibrium constants, the pe and the pH together provide a theoretical energy characterization of the redox system at equilibrium. Using a technique for slowly adding electrons to a soil suspension, Lindsay and Sadiq[3] demonstrated a series of pe + pH buffering regions related to the electron input and associated with specific mineral transformations taking place during reduction of the soil.

If their sum depicts an equilibrium constant, we can see that pe and pH are on opposite ends of a seesaw. One goes up; the other goes down. But the soil system is not a simple seesaw. The log in the middle keeps rolling.

For example, adding a source of electrons such as barnyard manure to a soil should lower the soil pe, and the lowered pe will favor oxidation. As oxidation takes place, pe will tend to rise. The tendency of the pH end of the seesaw to fall as pe rises, however, is counteracted by the release of basic cations from the manure as it decomposes. The pH rises faster than the pe, and this exerts a tendency to lower the pe end of the seesaw and increases the tendency of the manure to be oxidized at a faster rate. As oxidized species replace reduced ones (e.g., HNO_3 forms from NH_3), the pH tendency will be downward and the pe should go up, indicating an increasing tendency of the soil to become reduced.

Practically speaking, liming a soil can be expected to encourage oxidation, provided there is sufficient input of O_2. Liming will increase pH and lower pe. As oxidation of reduced species proceeds, the pe will rise and the pH will begin a downward trend. Thus, the soil disturbed by additions of acid or base has a proclivity toward adjusting itself to maintain a constancy of pe + pH. Lindsay and Sadiq[3] demonstrated that the sums of measured pe and pH remained constant over a range of values established by equilibrating suspensions of a sandy loam with increasing levels of HCl or NaOH.

REDOX SPECIES IN AERATED SOILS

pe-pH Diagrams

Constructing pe vs. pH diagrams using calculated equilibrium constants and assumed or measured equilibrium concentrations can be useful in comparing and thinking about possible redox reactions in aerobic soils.

[3] Lindsay, W., and M. Sadiq. Eleventh Int. Congr. Soil Sci. Oral presentation. Edmonton, Canada, Abstracts Vol. 1:138.

Obviously, we must be cautious in applying equilibrium diagrams to non-equilibrium soil systems. Still, they can help us visualize energy relationships between redox species and boundaries and patterns in the redox network. To illustrate Fig. 1 and 2 emphasize redox half-reactions we might expect in aerobic soils. The Fe line is included for reference to show the lower boundary of the aerobic zone. Carbon dioxide and N_2 reduction, though well below the Fe boundary, are included because they occur in aerobic systems.

If thermodynamic equilibrium prevails, the pe-pH region above any given line should favor the presence of the oxidized form of the couple; the region below the line, the reduced form. In the pe region above the Fe equilibrium line, the oxidized member of a couple can be thought of as "easily" reduced to the reduced member of the couple. Below the Fe line, the tendency is toward oxidation. Ferrous Fe is easily oxidized; oxidized Mn is easily reduced.

Thermodynamic diagrams deal only with energy relationships and not with probabilities or rates of reactions. For example, the diagram in Fig. 2 shows as a possible reaction the reduction of Mn coupled to the oxidation of Cr. Experience demonstrates that it is indeed a probable one (4). From the diagrams (Fig. 1 and 2), oxidation of Cr by NO_3^- might also appear to be possible at pH's above 6. But there is no evidence that this happens in soil.

In terms of energy, reduction of N_2 to NH_4^+, and CO_2 to glucose (Fig. 1), appear to be tough reductions down in the region where oxidation is "easy." We know, of course, that N-fixing microorganisms carry out the first reactions and all green plants the second. Paradoxically, both reduction reactions occur in aerobic systems. Both reactions use huge amounts of energy. Fixation of N requires considerable O_2 in order to mobilize metabolic energy for that remarkable reduction.

Oxidized species above O_2, e.g., N_2O, H_2O_2, and O_2^- (superoxide free radical), should be very "easily" reduced in soils and therefore will be powerful oxidizing agents. It should be easier to reduce N_2O to N_2 than NO_2^- to N_2O and easier to reduce NO_2^- to N_2O than NO_3^- to NO_2^-. Nitrous oxide and NO_3^- might both form from NO_2^-, N_2O from its reduction and NO_3^- from its oxidation. Thus, NO_2^- should be unstable and not accumulate in soils (44). In a general way, these theoretical suppositions are all suggestive of the actual soil behavior. They also remind us that equilibrium between the atmosphere and the soil does not exist. If it did, NO_3^- might accumulate in the soil until atmospheric O_2 became used up.

Quite low concentrations of atmospheric or dissolved O_2 can prevent anaerobiosis in field soils or natural waters, so long as input of O_2 equals or exceeds its depletion rate. This observation is reinforced by pe-pH diagrams, which are not much affected by large changes in O_2. In Fig. 2, equilibrium soil air that is 1% of atmospheric air gives an O_2/H_2O line that is only slightly below the 100% atmospheric equilibrium line. Practically, of course, 1% is so low that a very small amount of microbial or root respiration can quickly eliminate it, if soil pores are plugged with water.

Greenwood (23) demonstrated a changeover from aerobic to anaerobic metabolism in widely different soils when the O_2 became less than the

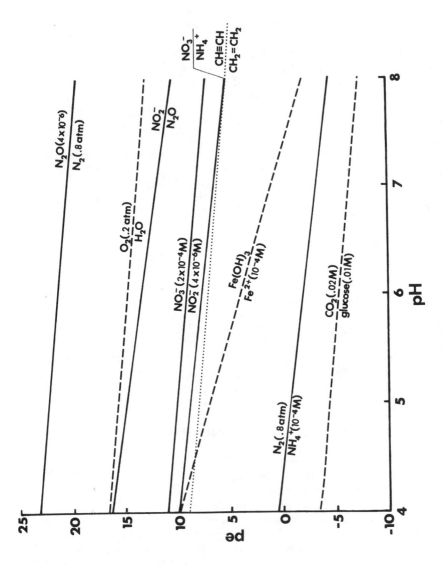

Fig. 1. Stability lines between oxidized and reduced species for several redox couples. Activities of species are designated if not equal within a couple.

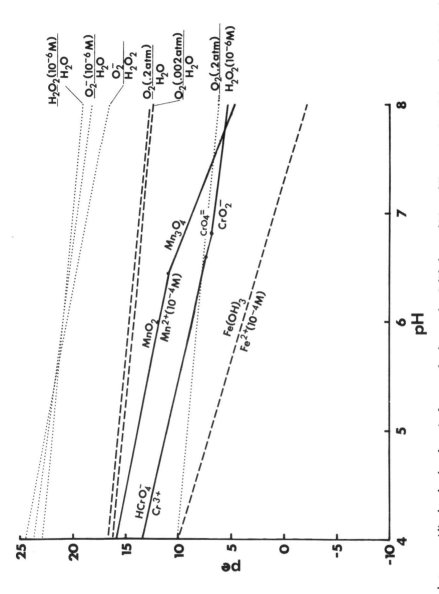

Fig. 2. Stability lines between oxidized and reduced species for several redox couples. Solid phases of Mn and Fe are at unit activity, and activities of other species are designated if not equal within a couple.

1% level. He concluded that since aerobic metabolism was not inhibited by lack of O_2 until it was very low, kinds of metabolic products, aerobic or anaerobic, could be used as accurate indicators of O_2 presence in soils.

To move the O_2/H_2O line down to the pe-pH level of the acetylene/ethylene line (still well above Fe) requires the absurdly low level of about 10^{-30} atm. of O_2. Based on assumptions of soil porosity, we calculated that this would be a partial pressure of 1 molecule of O_2 per hectare.

Poise

Poise relates to pe as buffer capacity relates to pH. Poise of the soil redox system is equivalent to the quantity of electrons required to change the pe by a given amount. The electrons are in the form of reduced substances that would donate them or oxidized substances that would accept them. Poise depends on amounts; pe depends on ratios of species in equilibrium.

Due to the irreversibility of the O_2 to H_2O half-reaction, aerobic soils are not poised by quantities of equivalent ratios of redox pairs in equilibrium. Nevertheless, aerated soils are mightily poised in the practical sense by atmospheric O_2 against rampant reduction and by soil organic matter against oxidation. Because of the irreversibility of half-reactions 1 and 2, adding reduced oxygen (H_2O) or oxidized carbon (CO_2) to a soil will not directly affect either pe or poise.

Continual inputs of both O_2 and reduced remains of plants and animals maintain a dynamic poise. Also contributing to poise are stable forms of soil humus rendered difficultly available to soil microorganisms by adsorption onto silicate clays, complexing by trivalent metals, or by phenolic glycosidic cross linkages within the molecular structure (21). Less easily reduced oxygen also is part of the poising picture in nitrates and Mn oxides. Oxidized Fe, of course, serves as a wall between the aerobic and the anaerobic states. The poising effect of reactive oxidized Fe must be overcome before a soil becomes totally anaerobic.

Waterlogging soils containing reactive fresh organic residues quickly brings anaerobic conditions if temperatures are suitable. However, short periods of O_2 exclusion will not result in anaerobiosis if the reduced C and N are mainly in the structure of meta-stable humus. Due to organic matter meta-stability and the insolubility of Mn oxides, some Mn may remain oxidized in waterlogged soils after O_2 has been depleted. There it poises the system against homogeneous anaerobiosis. Oxides of Mn may support aerobic microenvironments in a soil that otherwise is dominated by anaerobic species. Though far shorter lived, nitrates have similar effects. The flip side is represented by anaerobic microsites in aerated soils. These result from zones of poor aeration such as aggregate interiors or hot spots of highly reactive reduced organic substances.

Thus, poise or balance in aerobic soils is maintained at two dimensional levels. On the firing line are reactive carbonaceous compounds vs. reactive O_2. Backing up these readily available forms are meta-stable humus vs. meta-stable oxides of Mn and N, and back against the wall, standing between aerobic and anaerobic, are Fe oxides.

Low pe means high oxidation potential or tendency, and high pe means a high potential for reduction. But danger lurks in the additional step of translating high potential to mean high liklihood of change. This would be akin to saying, "I am so poor, I can only get richer." The water-logged O_2 deficient soil will continue to have a low pe as long as O_2 is excluded. A well-aerated soil will continue to have a high reduction potential as long as it draws on the inexhaustable atmosphere to poise it against reduction.

Reduction of Oxyanions

Oxidizing agents are divided into two categories, molecular elements such as O_2, Cl_2, and F_2, and oxyanions such as nitrate, chromate, and Mn oxide anions. Strong oxidizing agents in acid solutions, the oxyanions are, for kinetic reasons, ineffective in neutral solutions. The Walkley-Black (63) titration of soil organic C depends on oxidation by Cr^{6+} in H_2SO_4 solution. Yet Cr^{6+} can remain stable in a pH 6 high organic matter soil for months (4). Nitric acid will readily oxidize soil organic matter, but neutral salts of nitric acid would not be reduced by soil were it not for biological nitrate reductase systems.

Various aerobic heterotrophic soil bacteria that metabolize readily available organic C have the ability to use NO_3^-, NO_2^-, and N_2O as electron acceptors. Alexander (1) points out that in denitrification, "reducing power that normally is dissipated by reaction with O_2 is dissipated at the expense of nitrate." Manganese oxides are reduced similarly by soil microorganisms (1), but reduction is slowed by low solubilities of the oxides and by stabilities of organic complexes. In calcareous soils, reduction also is inhibited by high pH.

Reaction electron donating C compounds can directly reduce Mn oxides without intervention of microbial enzyme systems if pH is low. At soil pH's below 5.5, Mn^{2+} is the predominant Mn form (33). Though protons are consumed in the reduction of both Fe and Mn oxides, Mn reduction is different from that of Fe because it can frequently take place at high pe in the presence of O_2. Unlike that of Fe, availability of Mn to plants depends on its reducibility to the exchangeable and mobile divalent ion.

Oxidation of Mn

Like nitrification, soil oxidation of Mn requires a ready supply of O_2. Oxidation of Mn by a variety of microorganisms is easy to demonstrate in the petri dish, but it is difficult to prove that any single organism can oxidize Mn in soil. Unless a sterilized soil is inoculated with a cross-section of soil microorganisms capable of metabolizing the large amounts of available C compounds formed during sterilization, the oxidation of Mn will be suppressed by the reducing nature of the system. It is even difficult to prove that oxidation of Mn is directly biological because any agents inhibiting microorganisms are likely to inhibit non-biological oxidation processes as well.

However, there is agreement that microorganisms are vital in Mn oxidation (10, 11). Part of the difficulty in demonstrating oxidation by microorganisms is that the oxidation has been evaluated indirectly by the disappearance of Mn^{2+} (62). Determination of Cr oxidation by soil should be useful because it directly measures changes in oxidized soil Mn (4).

Wada et al. (62) concluded that Mn was oxidized slowly and non-biologically in a paddy soil on surfaces of tiny Mn oxide deposits, which they hypothesized were initially formed through rapid oxidation by microorganisms. Reduced Mn typically is re-oxidized in subsoils where higher pe and pH conditions prevail. Stains of Mn oxides occur deeper in the profile than precipitated Fe because Mn is less easily oxidized.

Adsorptive Role of Mn Oxides

Manganese oxides play a unique role in aerobic soils as highly adsorptive colloidal materials interacting with both soil mineral and organic fractions. These oxides are practically universal above pH 6 as mineral coatings and nodules on or near well-aerated surfaces. Their mineralogy is complex (45). Since they coat surfaces, they exert chemical influence far out of proportion to their total concentrations (28). Manganese oxides have extremely high surface areas and cation exchange capacities and appear to act as strong scavenging agents for heavy metals (12, 58). Charges are pH-dependent, with negative charges increasing markedly as pH is increased above 5 (46).

Jenne (28) proposed that hydrous oxides of Mn and Fe furnish the principal control on the fixation of Co, Ni, Cu, and Zn in soils and fresh water sediments. By selectively reducing a landfill soil, Suarez and Langmuir (59) showed that Mn-rich oxides had at least 10-fold higher heavy metal percentages than Fe-rich oxides, reflecting their greater coprecipitation potentials. McKenzie (37) found that addition of MnO_2 to Pb-contaminated soils lowered the uptake of Pb, Co, and to a lesser extent, Ni, by subterranean clover. The availabilities were related to the distribution of these elements between Mn and Fe oxides in the soils. According to Loganathan et al. (35), specific adsorption of Co and Zn by MnO_2, compared to Ca, was related to hydroxylation of the metal ions as pH was increased above 6.

Reduction of Mn oxides has dual effects on cation exchange in soils. Not only does the exchange surface disappear, but the newly formed Mn^{2+} enters into exchange competition with other cations. Activities of Ca, Mg, K, and adsorbed heavy metals may all increase as Mn reduces. Posselt and Anderson (46) found the adsorptive capacity of hydrous MnO_2 was considerably greater for Mn^{2+} than for any other divalent cation.

DRYING AND REWETTING EFFECTS

Redox behavior as affected by drying field moist soils and then later rewetting them deserves special mention and attention. An armchair soil scientist might deduce that air or oven-drying a soil sample would encourage equilibration with the atmosphere and thereby increase oxidation processes. Wrong. It has been shown again and again that drying makes soil systems more reducing.

Two sets of problems must be faced if moist soil is to be dried before or during the course of its study. First, drying may alter surface chemical characteristics so that the behavior of the dried sample, immediately after adding water to it, will differ from that of the continuously moist soil. Second, if the dried sample is to be remoistened during the study—for more than an hour or so—erratic, anomalous, or at least unpredictable results may be obtained while the material is in the process of changing back into its meta-stable moist state.

Drying in the laboratory or natural drying in the field topples the delicately poised meta-stable moist soil system causing instant changes. Reversing the process on rewetting occurs only slowly, requiring biological activity. The most obvious effect of drying is an increase in the solubility and reducing ability of soil organic matter (6, 52, 54, 56, 61). Surface acidity is increased (39), and Mn is reduced, becoming exchangeable and soluble (4, 19, 29, 30, 40). All of these effects increase with time of storage.

Increased solubility of organic matter probably results from the increased polarity of surface oriented water as drying proceeds. High surface tension of water may literally tear apart molecular and cell structure as contracting stress forces parallel to the surface approach thousands of atmospheres (13). Continued destruction of organic matter during storage may be analogous to deterioration of the stretched rubber band, noted to be much faster than that of the relaxed rubber band.

Raveh and Avnimelech (47) attributed broken organic structure associated with drying to breaking of hydrogen bonds, increased acidity to exposure of new acidic groups, and increased available energy to exposure of fresh organic surfaces. It seems likely that Mn reduction is associated with partial oxidation of the disrupted colloidal organic matrix.

CHARACTERIZING REDOX IN AERATED SOILS

Problems with Measured pe

Like pH, pe is not measured directly but is calculated by means of the Nernst equation from an electrical potential between electrodes inserted into the soil water suspension. The voltage observed between a Pt and a reference electrode, corrected for the effect of the reference, is multiplied by Faraday's constant and divided by 2.3RT to obtain the pe. In the case of pH, the meter does the calculating.

Theoretically, a reduced system will tend to lose electrons to the inert Pt electrode which will take on a negative charge (27). The redox potential and therefore the pe will be low. If the system becomes more oxidized, the electron activity will be lower, the charge on the Pt electrode will be more positive, and the pe measured will be higher. The glass electrode is sensitive to protons but is not directly affected by the electron activity. On the other hand, Pt is an all-purpose electrode. It will respond to pH as well as pe, or any other electrical potential present. In a system of constant pe, a one-unit decrease in pH will cause a one-unit increase in the apparent pe measured by the Pt electrode.

In theory, the presence of a few molecules of O_2 in an equilibrium solution should give the system a high pe. However, the influence of dissolved O_2 on the measured potential is far less than the high position of O_2 on the pe-pH diagram would lead us to think (20). Changes in the partial pressure of O_2 in equilibrium with dissolved O_2 are inadequately reflected in pe measurements (58). There also appears to be a tendency for dissolved O_2 to act on the Pt electrode so that the observed potential only partially reflects the oxidation states of other dissolved components (20). Neither Mn oxide nor NO_3^- has the expected quantitative effect on measured pe, Mn oxide because of its insolubility and NO_3^- because it is not electroactive (58).

Making a water suspension of soil has some effect in bringing a sample toward equilibrium within itself. The potential readings depend mainly on the relative proportions of various redox pairs that happen to be soluble in the water (20). Although solid sediments will influence the readings, their effects are not clear-cut.

Ponnamperuma (44) found soil potentials higher than in equilibrium solution before reduction but considerably lower than solution potentials after the soil was reduced. The pH of water on a clay surface is lower than in the bulk solution (39), and it seems likely that pe is higher on mineral surfaces than in bulk solution. However, close to organic matter surfaces or next to bacterial cells, pe should be lower than in the bulk solution (48).

Measured pe can't have precise theoretical significance unless we know which species are in equilibrium, the equations for the reactions involved, the values for the equilibrium constants, and most important of all we must know that there is indeed internal equilibrium in the system. Stumm and Morgan (58) discuss the theoretical reasons for discouraging experimental determination of pe in natural waters. The problems are still more serious in aerobic soil systems (7). A flooded soil certainly is closer to being a closed system than an aerated one. A working definition of an aerobic soil might be one in which pe measurements are of little or no theoretical value.

Bohn (8) pointed out the importance of clearly distinguishing between equilibrium electrode potentials (Eh), which are not amenable to determination in soils, and measured soil redox potentials (E_{Pt}). The latter are nonequilibrium mixed potentials. He stressed that their measurement is qualitative and their utility is in their interpretation by a knowledgeable observer. Often this distinction is rather fuzzily dealt with in the literature.

Empirical pe

Since the major redox components of aerobic soils have either ambiguous or small effects on measured potentials, one might question whether results of measurements in such soils are a waste of time at best or misleading at worst. Certainly wariness should be the watchword. But a measurement that is relatively easy to make, that is reproducible and that appears at least to qualitatively reflect redox conditions in a soil sample, should be worth the trouble. Its value is in comparing one soil sample with another, one treatment with another, or one time period with another.

We suggest that the term empirical pe, or EMpe, be used to designate an experimentally obtained value. Designation of the measured pe as empirical could help remind us that the value is not quantitatively precise and that we should not use it for thermodynamic interpretations.

Since not all redox half-reactions in soils consume or liberate one proton for each electron, it is not possible to assign a one to one relationship between pH and pe units in order to correct EMpe to a standard pH. Moreover, if a measured potential is to serve as an empirical mirror of soil redox status, there is little logic in trying to correct it for proton activity. Measured pH evaluates proton level separately; EMpe reflects both electron and proton levels, and both are related to redox status. Perhaps EMpe + pH will prove to be a more useful parameter than either measurement alone[3].

EMpe METHOD

In embarking on an electropotential determination, remember that two unlike electrodes always show a potential difference when placed into dirty water. Interpreting the reading is the point of the exercise.

a. Attach a bright Pt electrode to the plus terminal (in place of the glass electrode) of a pH meter with a millivolt scale and a saturated calomel electrode to the negative terminal. To make a Pt electrode, fuse both ends of a length of Pt wire into a glass tube so that a loop containing 3 or 4 cm protrudes, add a few drops of Hg to the inside of the tube for electrical contact between the wire and the lead removed from a discarded glass electrode.

b. Before each reading, rinse the Pt electrode (not the reference electrode) in a 1/1 6 N HCl/liquid detergent solution followed by 10% H_2O_2 and then thoroughly rinse with distilled water (55). Clean occasionally in aqua regia.

c. Adjust the potentiometer to read +219 mv. when the electrodes are in a pH 4 suspension of quinhydrone in 0.1 M K acid phthalate. This reading is equivalent to an Eh of 463 mv. and a pe of 7.85 at 25 C.

d. Add 30 ml of 0.01 M $CaCl_2$ to 10 g of soil (dry wt. basis) in a beaker or plastic cup, stir until soil and solution are well-mixed, and let stand for 20 to 30 min with occasional swirling.

e. Insert electrodes so that the reference electrode is in the upper half of the supernatant solution and the Pt electrode is near the bottom of the suspension. Swirl for a few seconds, let stand for at least 5 min, and without jiggling or touching the cup, read Eobs in mv. Measure pH of the same suspension.

$$pe = \frac{Eh}{2.3RT/F} \qquad EMpe = \frac{Eobs + 244 \text{ for satd. cal.}}{59} \qquad [7]$$

RESULTS

The measured EMpe was decreased in each soil listed in Table 1 by the mild drying treatment of 40 C for 12 hours. It was lowered more by air drying in the light for 2 months or by drying at 105 C overnight.

The Typic Eutrochrept (Lordstown loam) samples were kept flooded in the EMpe measurement cups for 10 days at room temperature. (Additional cups were set up for Cr oxidation tests, to be discussed later). The samples that had been dried overnight before flooding became anaerobic by odor beginning the third day and showed a positive Prussian blue test [acidified $K_3Fe(CN)_6$] for Fe^{2+} the fourth day. The soils that had been stored in the field moist state did not show evidence of anaerobiosis even after 30 days of flooding in open cups. The EMpe's (Table 2) reflected the various anaerobic and aerobic conditions.

Other Empirical Methods

Cr OXIDATION AND REDUCTION

Soils containing Mn^{3+} or Mn^{4+} oxides will oxidize Cr^{3+} to the Cr^{6+} chromate form, with the Mn acting as the electron acceptor (4). A fresh moist aerobic field sample with a pH above 6 typically will oxidize Cr (4). Thus, oxidation of Cr serves not only as a test for reactive oxidized Mn but

Table 1. Empirical pe's measured in five soil samples as affected by different drying treatments.

	Empirical pe				
	Typic Eutrochrept Al, loam	Aeric Haplaquept Ap, silt loam	Aquic Udorthent Al, lfs	Glossaquic Hapludalf Ap, clay	Typic Fragiorthod B2ir, loam
Field moist	9.2	10.5	8.9	11.3	12.1
Dried 40 C, 12 hours	8.8	9.9	9.3	10.8	9.9
Dried, greenhouse, 2 mo.	7.9	9.3	8.6	10.0	9.2
Dried, 105 C, 12 hours	8.4	9.5	8.2	9.8	8.6

Table 2. EMpe's and Cr oxidized by Lordstown loam Al (Typic Eutrochrept) as affected by drying treatments and flooding time.

	Time flooded				
	1 hour	1 day	3 days	6 days	10 days
	Empirical pe				
Field moist soil	9.2	9.2	9.4	8.8	8.9
Dried 40 C, 12 hours	8.8	8.3	6.8	2.1	1.8
	Cr oxidized				
	ppm				
Field moist soil	22	22	22	18	20
Dried 40 C, 12 hours	9.4	2.8	0	0	0

also as a means for identifying highly oxidized soils and for characterizing redox changes that occur when such soils are dried and stored and then are remoistened and incubated. The Cr reduction test measures easily oxidized soil organic matter such as that made soluble upon drying soils.

Oxidizing Test—Shake 2.5 g of soil (dry weight basis) 15 min with 25 ml 10^{-3} M $CrCl_3$, add 0.25 ml 1 M $KH_2PO_4 \cdot K_2HPO_4$, shake 30 sec more, and filter or centrifuge. Determine Cr^{6+} by adding 1 ml azide reagent to 8 ml of extract, mix, and let stand 20 min and compare color with that in standards at 540 nm. (Prepare reagent by adding 120 ml of 85% H_3PO_4, diluted with 280 ml of distilled water, to 0.4 g of s-diphenyl carbazide dissolved in 100 ml 95% ethanol.) Read absorbence of the original extract at 345 nm for organic matter color.

Reducing Test—Shake 2.5 g soil 1¼ hour with 25 ml of 3 ppm Cr^{6+} (0.29 M $K_2Cr_2O_7$) in 0.01 M H_3PO_4, filter or centrifuge and determine Cr not reduced in the extract. Compare organic colors by 345 nm absorbence.

Observations—The Cr oxidized does not measure nearly all of the oxidized Mn. Organic matter made soluble and Mn reduced by drying are closely related. Both effect Cr oxidation. Less Cr is oxidized after drying, and more of the Cr that is oxidized is reduced during the test by the increased soluble organic matter. Thus, the test measures net oxidation.

Leaching the Lordstown loam samples with 0.01 M KH_2PO_4 to remove soluble organic matter before running the Cr oxidation test increased net Cr oxidation (Fig. 3). In the sample dried at 105 C, there was no net oxidation until after leaching. Drying at 40 C for 100 hours was more severe in terms of reducing Mn than 105 C for 18 hours, but 105 C was more severe in releasing reducing organic compounds. Both Cr oxidizing and reducing tests respond to changes in availability of easily oxidized organic matter related to drying or rewetting soils.

Table 2 shows that flooding of the dried soil lowered net Cr oxidation drastically the first day, and it eliminated it altogether by 3 days; but in the soil stored in the field moist state, oxidizing ability was not significantly affected by flooding in open cups for 10 days.

Net Cr oxidized by Lordstown loam decreased as the amount of soluble orgnic matter (345 nm absorbence) in the test solution extract increased (Fig. 4). The color increase was related to severity of drying, with points from the lowest absorbence to the highest represented in order by soil at field moisture, soil dried overnight in the dark, and soil dried in the greenhouse in layers of decreasing thicknesses from 50 mm to 1 mm.

SOLUBLE COLOR

The amber color of a soil extract is another redox parameter. The increase in solubility of organic matter after drying (see IV) may be demonstrated by extract color or absorbence of short wave lengths of light or UV. Figure 5 shows increases in the color of 10/1 water extracts of the Lordstown loam as samples were allowed to dry from 15% moisture in humid-

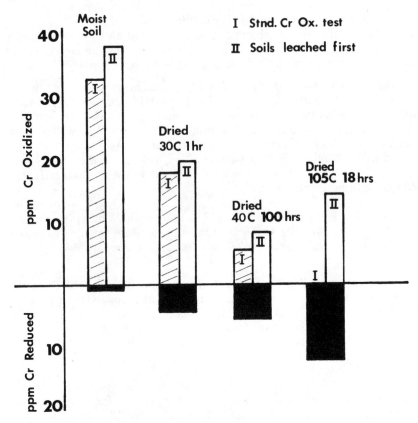

Fig. 3. Differences in oxidation of Cr by Lordstown loam samples unleached and after leaching with 0.01 M KH$_2$PO$_4$ to remove soluble organic matter. Chromium reduced by the reduction test also is shown.

ity jars over varying concentrations of H$_2$SO$_4$. The color disappears after a time in remoistened soils, apparently through microbial decomposition.

Drying, of course, wreaks havoc among microbial populations, killing millions of cells. Their remains undoubtedly contribute to the soluble organic matter (36), the increased reducing ability of which becomes evident when the soil is remoistened. Increased activities of microorganisms following rewetting of soils dried for storage or dried naturally in the field have been noted in many studies (5, 32, 54, 61, 56). Flooding soils after drying intensifies development of anaerobic conditions (Table 2). Skyring and Thompson (52) used rate of denitrification after flooding to estimate reduced organic matter made more available by soil drying. Air dry storage time was correlated with nitrification rate after rewetting soils (41). The influence of cultivation on soil drying and subsequent increase in N availability has been reported a number of times (2, 5, 32).

Twenty-five successive dryings at 80 C followed by 25 leachings with distilled water removed totals of 7,340 ppm C oxidizable by the Walkley-Black method (63) and 580 ppm Kjeldahl N from an Aquic Fragiorthod

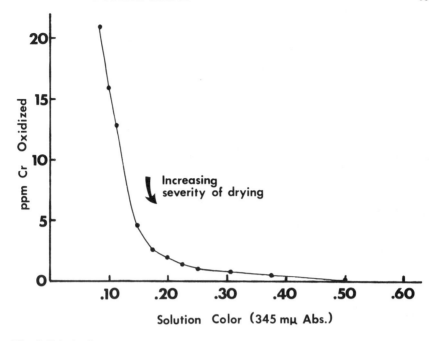

Fig. 4. Relation between net Cr oxidized and 345 nm absorbence in the Cr oxidizing test solution as drying severity was increased. Points from the top represent field moist, dried overnight in the dark, and dried in the greenhouse for 2 months in layers of decreasing thickness from 50 to 1 mm.

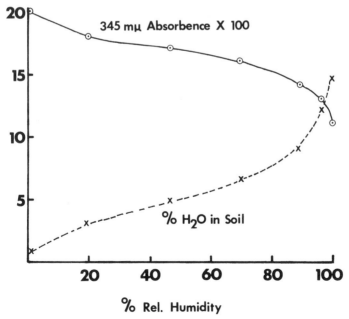

Fig. 5. Absorbence by extracted amber-colored water-soluble organic matter as related to soil drying for 2 weeks with humidity controlled by H_2SO_4.

Ap and 7,960 ppm C and 960 ppm N from a Typic Eutrochrept A1. In comparison, only trace amounts of C and N were removed after the first two leachings from soils kept moist.

On rewetting, decomposition of the organic matter made soluble by drying a soil will mineralize N. Assuming that C/N is fairly constant in the colored soluble organic matter, it follows that color of extracts of dried soil samples could predict comparative amounts of N that would be rapidly mineralized on rewetting them. Fox and Piekielek (17) found that UV absorption at 260 nm by 0.01 M NaHCO$_3$ soil extracts of dried samples was as well correlated with N-supplying capabilities of test soils in the field as other extractable N indexes.

Amadon measured 345 nm absorbence in extracts obtained from 67 Vermont soils by 15-min shaking of 2.5 g of dry soil with 25 ml of 0.01 M KH$_2$HPO$_4$ at pH 7.2 (J. F. Amadon, 1979. Soil organic matter characterization and N mineralization. M. Sci. Thesis. Univ. of Vermont). He found fair correlation between absorbence, soil organic matter, and total N. Correlations were best with recently dried samples.

Natural drying in the field also appears to affect N availability. Notice in Table 3 the differences in N mineralization under a ryegrass (*Lolium perenne*) sod in which the soil surface is relatively shaded from the sun compared with that under corn (*Zea mays*) in 90-cm rows, cultivated once (Magdoff, F. R., and J. F. Amadon. 1979. Unpublished results). The soil between the rows was subjected to considerable drying by the sun while the crop was young. The six-fold higher nitrate-N in the soil profile under corn was reflected in comparative yield response to applied N. Ryegrass yield was increased four times relative to the check by sludge or inorganic N, but yield of silage corn was not increased by additional N.

At a different site, lowered Cr oxidation tests were found in soil dried by the sun at the ground surface, compared with soil 2 to 4 cm below the surface or soil shaded by vegetation (Table 4). The Cr oxidized ranged from 0.6 ppm in a dry surface sample to 37.6 ppm under the shade of trees.

Soil redox effects of factors such as sunlight, wind, ground cover, tillage, non-tillage, mulching, shading, and other practices within the management sphere are worthy of further study.

Table 3. Nitrate N in the profile of Nellis silt loam (Typic Eutrochrept) on 26 July 1978 under ryegrass sod vs. 90-cm rows of cultivated corn.

Depth	Nitrate N, ppm of soil wt.	
	Ryegrass sod	Corn in 90-cm rows
cm		
0–15	3.2	27.6
15–30	4.7	17.0
30–60	2.0	23.4
60–90	0.9	3.1
90–120	1.1	6.1
Total to 120	11.9	77.2

Table 4. Soil moisture content and Cr oxidized by Hinesburg loamy fine sand (Entic Haplorthod) on 20 May 1979, after 1 week of dry, sunny weather.

Soil location	Depth	H_2O	Cr oxidized
	cm	%	ppm
Fallowed area	0–2	0.1	0.6
	2–4	7.9	3.9
Between crop rows	0–2	0.7	1.1
	2–4	14.3	5.4
Under grass sod	0–2	4.0	14.0
	2–4	5.6	8.3
Shade of trees	0–2	22.0	34.0
	2–4	23.6	37.6

TEST FOR PHENOLS

Levels of highly reduced soluble organic matter can be determined by tests designed to measure phenols or tannins (16). Any reducing agent strong enough to reduce phosphotungstate and phosphomolybdate to give the blue-colored complex will give a positive test. Reduced Fe, for example, will give the color as will many organics that are not phenols. Still, the test appears to be a practical means for characterizing soluble humic and fulvic acid fragments and microbial by-products. Water extracts of dried soils give highly positive tests relative to extracts of well-aerated soils kept continuously moist for periods of time. Probably most of the soluble readily oxidizable organic substances in soils are phenolic compounds. Schnitzer and Levesque (49) showed that ESR peaks indicating free radical contents of NaOH extracts of humified peat were directly proportional to the concentrations of the "phenolics" in the extracts.

In a meta-stable moist aerobic soil, S, Fe, and Mn will be in oxidized forms and the highly reactive organic compounds will have been decomposed. Only the most stable soil humus remains to poise the soil against oxidation. There may be a shortage of available electrons in such a soil. Merkle (38) observed in 1955 that Fe, Mn, and Cu are generally adequate for plants in soils that receive annual supplements of organic tissues. Fresh additions of easily oxidizable organic compounds could be vital in maintaining availability of trace metals to plants and in preventing accumulation of nitrates, superoxides, and peroxides.

The importance of a balance between electron acceptors and electron donating species in the plant rhizosphere has been little studied. There are multitudes of redox related organic compounds in the soil that have potential nutritional or growth regulating significance. Perhaps redox buffers of the quinone/hydroquinone type are important to plant health. Procedures need to be developed for analyzing such compounds in soils.

In calcareous soils, where reduction of Mn to forms available to plants is inhibited, it seems likely that Mn is most available to roots that generate low pe in their rhizospheres. Bromfield (9) showed that water soluble substances from air-dried soil and washings from oat roots dissolved Mn oxides. Water extracts of moist-incubated samples did not solubilize Mn.

Cyclic reduction of Mn oxides, followed by re-oxidation, appears to be important in maintaining Mn (and perhaps Fe) in hydrous amorphous readily reducible forms. Their behavior in plant rhizospheres merits study.

DISMUTATION OF H_2O_2

Hydrogen peroxide is easily reduced to H_2O and thereby acts as a strong oxidizing agent (Fig. 2). It also can be oxidized to O_2, but since this occurs in the middle pe range, H_2O_2 is not a powerful reducer. The electron donor for its reduction may be one of its own oxygens. A compound that oxidizes and reduces itself is said to undergo dismutation. In the following equation, one oxygen gives up an electron to become oxidized; the other becomes reduced by accepting the electron:

$$H_2O_2 \rightarrow H_2O + \frac{1}{2} O_2 \qquad [8]$$

Catalases are enzymes that destroy H_2O_2 by catalyzing its dismutation. Usually rate of "catalase" activity is assayed by manometrically measuring the amount of O_2 given off when a soil sample is treated with H_2O_2 (31, 51). Other procedures measure quantity of H_2O_2 reacted by titrating the residual (51). In an acid medium, oxidized Mn will oxidize H_2O_2; in an alkaline medium, Mn oxides, especially recently oxidized amorphous ones, will both oxidize and reduce H_2O_2; that is, they will catalyze its dismutation. Thereby they contribute to the so-called "catalase" activity. Ferric hydroxide also will catalyze the dismutation of H_2O_2, but not as fast as Mn oxides.

Since a small amount of catalyst will act upon a large amount of substrate, rate of dismutation is a better indication of quantities and activities of catalytic substances present than is amount of substrate dismutated. For whatever it tells us, measurement of the rate of H_2O_2 dismutation by a soil is an easy measurement to make, and it is fun to watch the bubbles rise.

Rate of dismutation may be evaluated by adding 5 ml of 0.5 M H_2O_2 solution to 2.5 g of soil and clocking the time required for the soil to evolve enough O_2 bubbles to displace 24 ml of H_2O (1 mmole of O_2). A good Cr oxidizing moist soil will dismutate 50 to 100 moles H_2O_2/kg/hour.

A high rate of O_2 evolution appears to indicate a highly oxidized soil. It may be yet another way to evaluate the quantity of reactive oxidized Mn in a soil, but it measures, in addition, activities of other dismutation catalysts, including real catalase. "Catalase" activity is high in organic matter surface horizons in neutral pH soils and is associated with presence of vegetation (51).

The free radical, superoxide (O_2^-) acts as a super reducing agent when it is oxidized to O_2. It is a super oxidizing agent when it is reduced to H_2O_2 or H_2O (Fig. 2). The enzymatic dismutation of O_2^- to H_2O_2 and O_2 and of H_2O_2 to O_2 and H_2O in the cells of aerobic organisms prevents the formation of the bio-destructive hydroxyl free radical, making aerobic life possible (18, 24).

REDUCED Mn

A soil test for available Mn also is a test for reduced and reducible Mn and therefore is a redox characterization of a soil. An adequate Mn soil test should include exchangeable Mn plus the portion of the oxidized Mn that is likely to become reduced and usable by plants (33). Most standardized tests involve drying the soil sample prior to determination of Mn^{2+} extractable by water, neutral salts, or chelating agents. Thus, the Mn^{2+} measured includes that already in the reduced state before the soil was dried plus that reduced during drying by readily oxidizable organic matter, and in addition, Mn reduced during the extraction process.

In 42 out of 50 Vermont corn soils extracted with $CaCl_2$, there was no exchangeable Mn present before air drying (4). The mean value of exchangeable Mn^{2+} in the 50 soils increased eight-fold on air drying.

Extraction of a soil with hydroquinone reduces more Mn than simple drying and removes the Mn component termed "easily reducible" (33, 50). Hydroxylamine extracts still more of the "easily reducible" Mn (15). Removal of Mn^{2+} from the solution phase by extraction with a chelating agent such as DTPA (34) will shift the equilibrium toward further reduction of solid phase oxidized Mn. Amount extracted will depend on the activity of the solid phase, pH, pe, and availability of electron donors. Possibly Mn^{3+} also is complexed and extracted.

Formation of Mn^{2+} from Mn^{3+} may occur without an organic electron donor by a dismutation reaction such as the following (14, 26):

$$2\,Mn^{3+} + 2\,H_2O = Mn^{2+} + MnO_2 + 4H^+ \qquad [9]$$

The reverse reaction is favored by pyrophosphate, which complexes Mn^{3+} (26), and in the case of hydrated colloidal MnO_2 by EDTA (25) and probably other organic complexing agents.

NITRATES AND NITRITES

The presence of NO_3^- in a soil is indicative of high pe, and the quantity of NO_3^- contributes to poise against reduction. Denitrification reflects lowered pe at the site of reduction resulting largely from high levels of organic matter availble for oxidation rather than by oxygen deficiency (43, 52). A similar statement could be made about ethylene accumulation as indicative of readily oxidizable organic matter (53).

Changes in NO_3^- level inside a buried polyethylene bag can reflect differences in soil moisture tension and natural drainage of the soil surrounding the bag (3). The buried bag approach was suggested as an integrated quantity measurement for making "aeration status" comparisons of soils in the field.

Observing the fate of highly reactive nitrites added to a soil could provide a sensitive means of evaluating direction and rates of redox changes occurring.

BENZIDINE

The high pe redox indicator benzidine has been long considered a specific test for oxidized soil Mn (9, 33, 62). When a few drops of 1% benzidine in 2 N acetic acid are added to a pinch of moist soil on a spot plate, dark blue points and zones show up almost immediately. It is probably correct to consider benzidine as fairly specific in staining of soil Mn oxides, even though it gives blue color with other high pe substances such as Cr^{6+} and chlorinated tap water, neither likely to be present in most soils. It does not turn blue with nitrates, nitrites or H_2O_2.

One can gain a feel for soil Mn by spending some time examining benzidine-treated soil samples under the low power (10X) of a binocular microscope (in the hood since benzidine is carcinogenic). The Mn oxides usually occur as coatings on peds or sand particles, much like the macro stains that can be observed without benzidine in many soils. Close association with organic matter often is apparent. Extreme lack of soil homogeniety is the most striking observation. Successive drying of the Lordstown loam samples at 80 C followed by leaching with water removed all of the oxidized Mn detectable by benzidine.

SNIFF TEST

To the talented and trained nose, soils offer many clues related to both pe and pH. A classic example is the distinctive odor of freshly plowed ground as indicative of actinomycete activity (48) in soils with high pe and near-neutral pH. It takes no rare ability to recognize odoriferous outputs of sulfides, amines, scatol, and other products of anaerobic putrefaction. But accurate olfactory evaluation of aerobic soil systems requires a gifted nose and years of nasal nurture, tutelage, and careful calibration. The truffle snuffling pig is an example of art perfected through practice.

If we become too dependent on spectrophotometers and potentiometers, we are in danger of losing our special ability to sniff out a soil and make use of common scents to know when the pe and pH are both just right. Perhaps it is already too late.

LITERATURE CITED

1. Alexander, M. 1977. Introduction to soil microbiology, 2nd ed. John Wiley & Sons, New York.
2. Allison, F. E. 1973. Soil organic matter and its role in crop production. Elsevier Scientific Pub., New York.
3. Bartlett, R. J. 1965. A biological method for studying aeration status of soil in situ. Soil Sci. 100:403–408.
4. ————, and B. R. James. 1979. Behavior of chromium in soils: III. Oxidation. J. Environ. Qual. 8:31–35.
5. Birch, H. F. 1958. The effect of soil drying on humus decomposition and nitrogen availability. Plant Soil 10:9–13.
6. ————. 1964. Mineralization of plant N following alternate wet and dry conditions. Plant Soil 20:43–49.

7. Bohn, H. L. 1968. Electromotive force of inert electrodes in soil suspensions. Soil Sci. Soc. Am. Proc. 32:211–215.

8. ————. 1971. Redox potentials. Soil Sci. 112:39–45.

9. Bromfield, S. M. 1958. The solution of γMnO_2 by substances released from soil and from the roots of oats and vetch in relation to manganese availability. Plant Soil 10:147–160.

10. ————. 1978. The effect of manganese-oxidizing bacteria and pH on the availability of manganous ions and manganese oxides to oats in nutrient solutions. Plant Soil 49:23–39.

11. ————, and D. J. David. 1976. Sorption and oxidation of manganous ions and reduction of manganese oxide by cell suspensions of a manganese oxidizing bacterium. Soil Biol. Biochem. 8:37–43.

12. Chao, T. T. 1972. Selective dissolution of manganese oxides from soils and sediments with acidified hydroxylamine hydrochloride. Soil Sci. Soc. Am. Proc. 36:764–768.

13. Cohen, A. L. 1973. Critical point drying. In M. A. Hyat (ed.) Principles and techniques of scanning electron microscopy. Van Nostrand Reinhold Co., New York.

14. Dion, H. G., and P. J. G. Mann. 1946. Three-valent Mn in soils. J. Agric. Sci. 36:239–245.

15. ————, ————, and S. G. Heintze. 1947. The "easily reducible" Mn of soils. J. Agric. Sci. 37:17–22.

16. Folin, O., and V. Ciocalteu. 1927. On tyrosine and tryptophane determinations in proteins. J. Biol. Chem. 73:627–649.

17. Fox, R. H., and W. P. Piekielek. 1978. A rapid method for estimating the N-supplying capability of a soil. Soil Sci. Soc. Am. J. 42:751–753.

18. Fridovich, I. 1975. Superoxide dismutases. Ann. Rev. Biochem. 44:147–159.

19. Fujimoto, C. K., and G. D. Sherman. 1945. The effect of drying, heating, and wetting on the level of exchangeable manganese in Hawaiian soils. Soil Sci. Soc. Am. Proc. 10:107–112.

20. Garrels, R. M., and C. L. Christ. 1965. Solutions, minerals, and equilibria. Harper and Row, New York.

21. Gray, T. R. G. 1976. The survival of vegetative microbes in soil. p. 327–364. In T. R. G. Gray and J. R. Postgate (ed.) The survival of vegetative microbes. Cambridge Univ. Press, Cambridge.

22. Greenland, D. J. 1962. Denitrification in some tropical soils. J. Agric. Sci. 58:227–233.

23. Greenwood, D. J. 1961. The effect of oxygen concentration on the decomposition of organic materials in soils. Plant Soil 14:360–376.

24. Halliwell, B. 1974. Superoxide dismutase, catalase, and glutathione peroxidase: solutions to the problems of living with oxygen. New Phytol. 73:1075–1086.

25. Heintze, S. G. 1957. Studies on soil. Mn. J. Soil Sci. 8:287–300.

26. ————, and P. J. G. Mann. 1949. Studies on soil manganese. J. Agric. Sci. 39:80–95.

27. Hesse, P. R. 1972. p. 437–459. A textbook of soil chemical analysis. Chem. Publ. Co., Inc., New York.

28. Jenne, E. A. 1968. Controls on Mn, Fe, Co, Ni, Cu, and Zn concentrations in soils and water: the significant role of hydrous Mn and Fe oxides. In R. F. Gould (ed.) Trace inorganics in waters. Advan. Chem. Series 73:337–389.

29. Kelley, W. P., and W. McGeorge. 1913. The effect of heat on Hawaiian soils. Hawaii Agric. Exp. Stn. Bull. 30.

30. Khanna, P. K., and B. Mishra. 1978. Behavior of Mn in some acid soils in western Germany in relation to pH and air-drying. Geoderma 20:289–297.

31. Kuprevich, V. F., and T. A. Shcherbakova. 1966. p. 150–194. In Soil enzymes. Translated from Russian, USDA and 1971 Ind. Nat. Sci. Doc. Ctr., New Delhi.

32. Lebedjantsev, A. N. 1924. Drying of soil as one of the natural factors in maintaining soil fertility. Soil Sci. 18:419–447.

33. Leeper, G. W. 1947. The forms and reactions of Mn in the soil. Soil Sci. 63:79–94.

34. Lindsay, W. L., and W. A. Norvell. 1978. Development of a DTPA soil test for Zn, Fe, Mn, and Cu. Soil Sci. Am. J. 42:421–428.

35. Loganathan, P., R. G. Burau, and D. W. Fuerstenau. 1977. Influence of pH on the sorption of Co^{2+}, Zn^{2+}, and Ca^{2+} by a hydrous manganese oxide. Soil Sci. Soc. Am. J. 41: 57–62.

36. Marumoto, T., H. Kai, T. Yoshida, and T. Harada. 1977. Chemical fractions of organic nitrogen in acid hydrolysates given from microbial cells and their cell wall substances and characterization of decomposable soil organic nitrogen due to drying. Soil Sci. Plant Nutr. 23:125–134.

37. McKenzie, R. M. 1978. The effect of two manganese dioxides on the uptake of lead, cobalt, nickel, copper, and zinc by subterranean clover. Aust. J. Soil Res. 16:209–214.

38. Merkle, F. G. 1955. Oxidation-reduction processes in soils. p. 200–218. In F. E. Bear (ed.) Chemistry of the soil. Reinhold Pub., New York.

39. Mortland, M. M., and K. V. Raman. 1968. Surface acidity of smectites in relation to hydration, exchangeable cation and structure. Clays Clay Min. 16:393–398.

40. Nelson, L. E. 1977. Changes in water soluble Mn due to soil sample preparation and storage. Comm. Soil Sci. Plant Anal. 8:479–487.

41. Nevo, Z., and J. Hagin. 1966. Changes occurring in soil samples during air-dry storage. Soil Sci. 102:157–160.

42. Nommik, H. 1956. Investigations on denitrification in soil. Acta Agric. Scand. VI 2:195–228.

43. Patrick, W. H., Jr. 1960. Nitrate reduction rates in a submerged soil as affected by redox potential. Trans 7th Int. Congr. Soil Sci. 2:494–500.

44. Ponnamperuma, F. N. 1972. The chemistry of submerged soils. Adv. Agron. 24:29–96.

45. ————, T. A. Loy, and E. M. Tianco. 1969. Redox equilibria in flooded soils: II. The Mn oxide systems. Soil Sci. 108:48–57.

46. Posselt, H. S., and F. J. Anderson. 1968. Cation sorption on colloidal hydrous manganese dioxide. Environ. Sci. Tech. 2:1087–1093.

47. Raveh, A., and Y. Avnimelech. 1978. The effect of drying on the colloidal properties and stability of humic compounds. Plant Soil 50:545–552.

48. Russell, E. W. 1973. Soil conditions and plant growth. Longman, London.

49. Schnitzer, M., and M. Levesque. 1979. Electron spin resonance as a guide to the degree of humification of peats. Soil Sci. 127:140–145.

50. Sherman, G. D., and P. M. Harmer. 1942. The manganous-manganic equilibrium in soils. Soil Sci. Soc. Am. Proc. 7:398–405.

51. Skujins, J. J. 1967. Enzymes in soils. p. 371–414. In A. D. McLaren and G. H. Peterson (ed.) Soil biochemistry. Marcel Dekker, Inc., New York.

52. Skyring, G. W., and J. P. Thompson. 1966. The availability of organic matter in dried and undried soil, estimated by an anaerobic respiration technique. Plant Soil 24:289–298.

53. Smith, K. A., and S. W. F. Restall. 1971. The occurrence of ethylene in an aerobic soil. J. Soil Sci. 22:430–443.

54. Sorensen, L. H. 1974. Rate of decomposition of organic matter in soil as influenced by repeated air drying-rewetting and repeated additions of organic material. Soil Biol. Biochem. 6:287–292.

55. Starkey, R. L., and K. M. Wight. 1945. Anaerobic corrosion of iron in soils. Am. Gas Assoc., New York.

56. Stevenson, I. L. 1956. Some observations on the microbial activity in remoistened air-dried soils. Plant Soil 8:170–182.

57. Stolzy, L. H., and H. Fluhler. 1978. Measurement and prediction of anaerobiasis. p. 363–426. In D. R. Nielsen and J. G. MacDonald (ed.) Nitrogen in the environment. Vol. 1, Academic Press, New York.

58. Stumm, W., and J. J. Morgan. 1970. Aquatic chemistry. Wiley-Interscience, New York.

59. Swarez, D. L., and D. Langmuir. 1976. Heavy metal relationship in a Pennsylvania soil. Geochimica Cosmochimica Acta. 40:589–598.

60. Van Breemen, N., and R. Brinkman. 1976. Chemical equilibria and soil formation. p. 163–164. *In* G. H. Bolt and M. G. M. Bruggenwert (ed.) Soil chemistry A. basic elements. Elsevier Scientific Pub., New York.

61. Van Schreven, D. A. 1967. The effect of intermittent drying and wetting of a calcareous soil on C and N mineralization. Plant Soil 26:14–32.

62. Wada, H., A. Seirzyosakol, M. Kimur, and Y. Takai. 1978. The process of manganese deposition in paddy soils. I. Soil Sci. Plant Nutr. 24:55–62.

63. Walkley, A., and I. A. Black. 1934. An examination of the Degtjareff method for determining soil organic matter and a proposed modification of the chromic acid titration method. Soil Sci. 37:29–38.

QUESTIONS AND ANSWERS

Q. Are you pessimistic about redox determination in soils because you question the reliability of electrode measurements of electron activity or do you feel there are problems related to particular theoretical couples?

A. I believe the question has to do with whether we worry about lack of theoretical significance of pe measurements or whether we have doubts that the electrodes respond to the differences that are really there. In the first place, I don't really feel pessimistic about making pe measurements. The electrodes do respond to real and significant differences between soils. I do not think we should feel discouraged if we can't find a theoretical interpretation that gives precise meaning to our results. This is particularly a problem in aerobic soils. My plea is that we should simply go ahead and make redox measurements in aerobic soils in order to compare soils and not worry if we can't attach precise theoretical meaning to the measurements.

Q. But to test whether or not the results have theoretical application, you must measure the quantity of both oxidized and reduced species.

A. Yes, and this may be difficult, perhaps impossible sometimes, because we don't even know which species are there. I'm saying we should go ahead and make a measurement, in spite of our ignorance, the way we do with pH. We can evaluate the effects of treatments on soils, and we can demonstrate that two soils are different. I agree that the more rigorous approach you talk about is extremely valuable when we know enough. I am trying to bridge the gap between the more complete understanding you refer to and the strictly empirical approach where we simply dump some sludge on a soil to see what happens.

Q. Have you found in the literature where they have used polarographic techniques to try to get at the redox pairs?

A. No we didn't get into that.

Q. I like the things that can be done with pe and pH, but by making the analogy between pe and pH we may imply there are free electrons in soils. I think this would be a mistake. Since we can express electron potential in terms of Eh, we aren't forced to make that kind of inference about free electrons.

A. I became attracted to the pe concept in reading Lindsay's work. I like the idea of thinking about pe as analogous in energy to pH. In the energy sense, the two are equivalent. There are no free electrons, of course. In adding reduction and oxidation half reactions, the electtrons must cancel out. The active electrons that we talk about are in the forms of reduced species. We talk a lot about hydrogen ions, but they are not free either. To me, Eh has become identified with reduced systems. By switching to pe, we recognize the entire redox spectrum.

CHAPTER 6

Eh and pH Measurement in Menfro and Mexico Soils[1]

R. W. BLANCHAR AND C. E. MARSHALL[2]

INTRODUCTION

Reactions which occur in soils are often difficult to discern due to the number of uncontrolled variables in a soil system. Some variables are controlled when soils are removed from their natural setting and placed in controlled environments. The environment then can be adjusted to give

---

[1] Contribution from the Missouri Agric. Exp. Stn. Journal series no. 8386. Received and approved by the director.
[2] Professor of agronomy, and professor emeritus of agronomy, respectively. Dep. of Agronomy, University of Missouri, Columbia, MO 65211.

maximum expression to the reaction in question. This approach has been useful, but it has not elucidated the role that each reaction plays in the chemical processes occurring in soil systems in their natural place. An experimental approach which appears very promising is to make key measurements on soils in their natural setting so that the rate and position of various chemical reactions can be predicted from data collected under controlled conditions.

Temperature, pressure, volume, and composition influence the direction, rate, and final condition of chemical reactions and are considered to "describe the system" at any point in time. Soil in place may be considered an open system with periodic exchanges of mass and energy from outside. The frequency, intensity, and duration of these exchanges to a large extent control the changes in composition which occur. Recently a great deal of attention has been paid to the use of oxidation potential (Eh), hydrogen ion activity expressed as (pH), and ion activity to predict the stability of compounds in soil.

The emphasis of this study is to examine the validity of Eh and pH measurements in a Menfro and Mexico soil. Experiments were carried out over a number of years. Some data have been reported elsewhere, but most are original research. The Menfro soil occupies the summit and upper shoulder positions of the deep loess hills bordering the Missouri River. The Menfro soil is a member of the fine-silty, mixed, mesic family of Typic Hapludalfs. The Mexico soils occur in summit positions of broad interstream divides in the loess covered Kansan till plain. The Mexico series is a fine, montmorillonitic mesic, Udollic Ochraqualf characterized by poor internal water drainage due to clay accumulation in the B horizons. These soils were chosen because the Menfro represents soil with better internal drainage characteristics than the Mexico. Also comparisons of Eh, pH, iron and manganese could be made using these soils.

Any compound which could be visualized to react with water and either add or remove hydrogen ions or electrons may be depicted on Eh-pH diagrams. Interpretation of Eh or pH should be based on the understanding that they are hypothetical rather than measured. However, under certain conditions theoretical and measured Eh and pH are quantitatively related. An example of such a diagram is the one prepared by Feagley (1979) for manganese which shows the limits of Eh and pH for natural systems determined by Bass Becking et al. (1960) (Fig. 1). The interpretation which can be made is that if the Eh and pH pertinent to the reactions were known the relative stability of the various components shown in the diagram could be predicted. The question which needs to be answered is what estimate of Eh and pH must be made in order to be applicable to those used to establish the stability fields shown in Fig. 1.

SOME OBSERVATIONS ON THE USE OF Eh and pH IN SOILS

Ponnamperuma et al. (1967) concluded that the soil solution of flooded soils is a thermodynamically meaningful phase. This conclusion was based upon the observation that measured iron in solution and that computed from Eh and pH measurements agreed. Collins and Buol (1970)

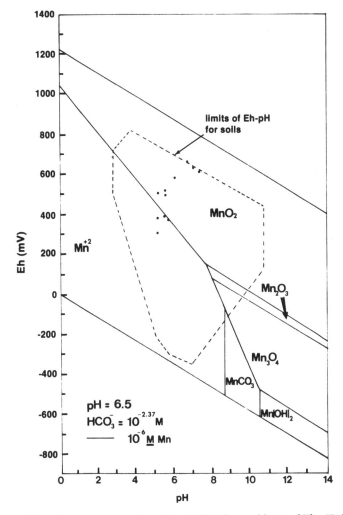

Fig. 1. Stability of manganese at various Eh and pH values and limits of Eh-pH observed for soils (Baas Becking et al., 1966). Computed for freshly precipitated solids using E° values of Ponnamperuma et al., 1969.

concluded that the validity of Eh-pH stability diagrams for either iron or manganese stability in soils was demonstrated by their experiments. They also found that in mixtures of manganese and iron precipitates Eh and pH relationships were not quantitative for the manganese. Pasricha and Ponnamperuma (1976) in reviewing their work on submerged soils concluded that the qualitative changes in iron and manganese could be predicted from $Fe(OH)_3$, $Fe_3(OH)_8$, and $MnCO_3$ as solid phases along with measurement of PCO_2, pH and Eh. Gotoh and Patrick (1974) measured Eh, pH, and iron in solution in waterlogged soil and concluded that ferrous ion in the soil solution would not exceed that predicted by amorphous ferric hydroxide or goethite.

Blanchar and Scrivner (1972) postulated that the solubility of ferric hydroxides in a Menfro soil might vary with depth. The Eh values they measured were in the presence of about 10^{-4} molar iron and were considered reasonable estimates from which to compute the $Fe(OH)_3$ ion product. These observations indicated that the iron oxides in the Menfro soil had solubility characteristics similar to goethite. The values of pFe + 3 pOH in the Mexico soil varied around 38.6 when 0.01 N HCl was added and around 40.6 when 0.1N HCl was added and the equilibrating solution contained more iron (Kao and Blanchar, 1973). Langmuir and Whittemore (1971) measured pH, Eh, and total dissolved iron in 24 well water samples where the ferrous ion content varied from 10^{-3} to $10^{-5.4}$M and the ferric oxyhydroxide content varied from 10^{-4} to $10^{-7.4}$ M. They found pFe + 3 pOH values of 39 to 43 and concluded that ferric oxyhydroxides in natural waters are composed of amorphous ferric hydroxide and goethite. These conclusions are consistent with the nature of iron hydroxides present in Menfro and Mexico soils.

Feagley (1979) measured the manganese concentration, pH, Eh, and electrical conductivity of solutions equilibrated with Menfro surface and subsurface horizon samples. He concluded that a MnO_2 form similar to birnesite probably existed in the Menfro soil. He estimated the $E°$ value for Eq. [1] to be 1060 mV.

$$Eh = E° - 120\,pH - 30 \log (Mn^{+2}) \qquad [1]$$

The $E°$ value reported for pyrolusite is 1,228 mV and that used by Ponnamperuma et al. (1969) and Yamane (1973) for freshly precipitate MnO_2-like material was 880 mV.

Feagley (1979) measured Eh and pH in a Menfro soil in place and could not predict total dissolved manganese from these measurements. Feagley pointed out that if at constant pH the Eh in Eq. [1] varied by 50 mV the manganese level would be predicted to change 46 fold. Feagley (1979) estimated that the average variation of properly functioning platinum electrodes buried in the Menfro soil was 50 ± 10 mV, large enough to vary manganese content widely.

Eh, measured with a platinum electrode, has been an unreliable estimate of the proportion of oxidized and reduced species in soils and waters (Morris and Stumm, 1967; Bohn, 1968, 1969, 1970; and Whitfield, 1969, 1974). The failure of Eh to be quantitative has been attributed to a lack of electro-activity of the couples measured, insufficient electrode current due to low concentrations, the formation of precipitates on the platinum surface, and lack of electron transfer mechanism between components of the system. A "mixed potential" may exist in natural systems when a transfer of electrons from one part to the other does not occur or is extremely slow. For instance hydrogen gas and ferric ions interact only through processes involving intermediary reactions and are considered to be poorly coupled.

Marshall (1977) emphasized that the striking appearance of colors in soil associated with heterogeneous distribution of iron and manganese could be explained by the factors contributing to their solubility and mobility. Relatively immobile iron and manganese require inputs of energy and microbial activity to reduce them to more mobile constituents in soil systems. Heterogeneity is due to surfaces, roots, organic fragments, and other factors which contribute to localized microbial activity and the development of reduction zones in soils. This heterogeneity is not indicated by the normal measurements of Eh and pH. However, this does not preclude the usefulness of these estimates for understanding how the system functions. The purpose of this study is to describe the Eh and pH and ion activity measurement for soils in place and evaluate interpretations which can be made from them.

MEASUREMENT OF SOIL Eh AND pH

Usually soil pH is measured by removing a sample, adding a solution, inserting glass and calomel electrodes, and measuring the output. For most in situ applications the hydrogen sensitive electrodes have been considered too fragile to insert and leave in soil as monitoring devices. The pH of 7.5, 15, 30, and 60 cm layers of a Mexico soil was measured throughout 1973 and 1974 and each layer varied between pH 5 and pH 7 (Hess and Blanchar, 1977). Variations in pH of this magnitude are sufficient to shift the position of many equilibria enough to change the relative stability of some minerals, however these pH changes do not represent as great a potential difference as those associated with Eh.

The Eh may be measured by removing a sample, adding a solution, inserting platinum (Pt) and calomel electrodes, and measuring output. In many cases Pt wires are buried in the soil in its natural setting and the output periodically measured. These in situ measurements are possible due to the physical stability of the Pt wire, but have been questioned by many investigators due to surface interactions between soil and the Pt wire. On the Mexico soil pH varied from 5 to 7 while the Eh varied about 600 mV. In terms of a potential shift, a 600 mV range for Eh represents about a 5 fold greater intensity than that associated with a shift of 2 pH units.

RELATIONSHIP OF MEASURED Eh
TO OXIDATION POTENTIAL

The Eh is an estimate of a system's oxidation potential. It also could be expressed as a partial oxygen pressure. The general relationship relating Eh, pH, and PO_2 for water is:

$$Eh = 1230 + 15 \log PO_2 - 59 \, pH. \qquad [2]$$

Thus in aerated soil the Eh should be fixed at any given pH since if PO_2 equals 0.2 atm, $Eh = 1230 - 59 \, pH$. This relationship has never been observed in natural environments. Bass Becking et al. (1960) reviewed hundreds of Eh and pH measurements in natural environments and the Eh measured is at least 200 mV less than predicted by Eq. [2]. Stumm and

Morgan (1970) discuss the amount of exchange current that a reaction must be capable of generating before it can be detected. They estimate that platinum electrodes with ordinary pH meters require an exchange current of 10^{-7} ampere when the O_2 pressure is 1 atmosphere. Thus the influence of PO_2 on Eh is not directly reflected in the potential measured with the platinum electrode. Sato (1960) has postulated that the formation of hydrogen peroxide as an intermediate is the rate limiting step responsible for the observed potential in oxygenated waters. Another explanation is that the Pt electrode in acidic solutions responds to the platinum oxide potential $Pt(OH) + e^- = Pt + OH^-$ rather than to PO_2.

In soil the measured potential is a mixed potential and may or may not reflect the position of either the oxygen or iron electron transfer reaction. Ferrous ion in the presence of dissolved O_2 gives an Eh reading which slowly drifts and is of little value in predicting the position of either reaction. Rapid electron transfer occurs between ferrous and ferric ions and the platinum electrode in acid under N_2 gas. The Eh measurement quantitatively represents the condition of this reaction. However, the following reactions of interest in soils such as organic oxidation-reduction couples and those for $NO_3^- - NO_2^- - NH_4^+$; $SO_4^= - H_2S$; and $CH_4 - CO_2$ have established reversible potentials, but react slowly with many other electron donors and acceptors in natural soil systems and are therefore considered to be "perched." The following experiments were done to resolve some questions associated with Eh measurements in soils.

Eh, pH, AND GAS PHASE VARIATION FOR A MEXICO Ap HORIZON SOIL SAMPLE

One hundred grams of soil and 200 ml of either water or 5% dextrose were added to the soil and allowed to stand for either 7 or 30 days. The Eh and pH were measured using the multiple electrode arrangement shown in Fig. 2. Two glass electrodes, two platinum and two calomel electrodes were connected to an automatic electrode switch which permitted the reading of potential output for 12.5 sec at 75 sec intervals or from any one electrode continuously. Various gases were bubbled into the system and the Eh and pH measured.

The measured Eh of a Mexico soil varied as the gas phase was changed from air, N_2, to H_2, O_2, or CO_2 (Fig. 3). The Eh + 59 pH values indicated that when H_2 gas was bubbled into the soil which had been equilibrated for 30 days with water, the potential was near zero as predicted if the Pt electrode was acting as a H_2, H_a^+ electrode. This was not true when the soil was equilibrated for only 7 days or when glucose was added. Where glucose was added the Eh measurement after 30 days was nearly independent of the gas phase, except when O_2 was bubbled into the system and an irreversible increase in Eh took place (Fig. 3). The iron content of these solutions was determined at 7, 15, and 30 days. With no glucose the iron content was 0.8, 1.8, and 4 μm/liter and with added glucose it was 604, 1,920, and 4,940 μM/liter. The independence of Eh to gas type in the glucose treatment is attributed to the buffering effect of the Fe^{+3}, Fe^{+2} in solution. The dependence of Eh on the gas phase in the soil

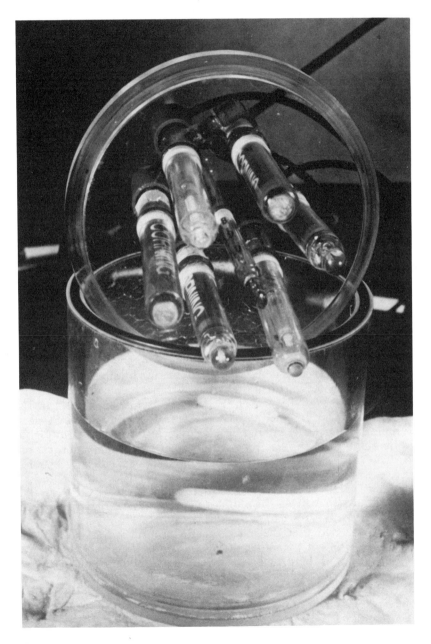

Fig. 2. Multiple electrode arrangement used to measure Eh and pH under various gases.

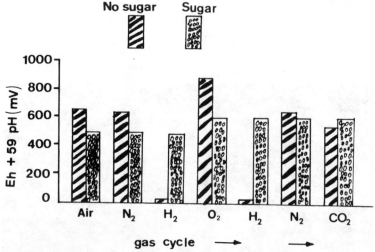

Fig. 3. The Eh of a Mexico soil after 30 days of submergence with either water or a 5% dextrose solution as influenced by various gases.

without glucose was attributed to the small amount of $Fe^{+3} - Fe^{+2}$ and other ionic buffers against Eh change in solution and the relatively greater response of the electrode to reactions involving either H_2 or O_2.

PERCHED HYDROGEN AND IRON COUPLES IN 1M HCl

The relative response of the platinum electrode to $Fe^{+2} - Fe^{+3}$ and $H_2 - H^+$ was measured in HCl systems. Four bright platinum electrodes (two button type and two wire type) were placed in the cell shown in Fig. 2. Hydrogen gas was continuously bubbled into the solution and the iron concentration varied. The Eh of the system was continuously recorded.

At an average activity of iron ($Fe_a^{+2} = Fe_a^{+3} = 1.33 \times 10^{-4}$) in 1N HCl and with H_2 gas bubbled into the system, the output of the Pt electrode could be drastically altered by stirring. The curve shown in Fig. 4 shows the response of the electrode with time after turning the stirrer on or off. The process was repeated many times and appeared reversible. When the stirrer was off H_2 bubbles were observed to form on the Pt surface and the Eh dropped to that predicted for the $H_2 - H^+$ couple. Turning on the stirrer broke the bubbles and the Eh reading was closer to that predicted for the iron couple.

It was assumed that the measured Eh of two couples which are "perched" at different potentials would be proportional to the current carried by each and their respective potentials. For conceptual purposes the proportion of current carried by each couple was considered proportional to the fraction of the Pt electrode covered by each. The equation derived has the same form as that given by Eisenman (1969) for ion selective electrodes:

$$Eh = (1-f)E^o_{H_2} + fE^o_{Fe} + (1-f)59 \log \frac{(H^+)}{(H_2)} + f59 \log \frac{(Fe^{+3})}{(Fe^{+2})} \qquad [3]$$

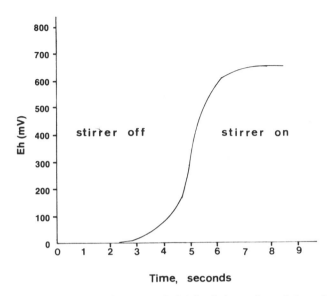

Fig. 4. The effect of stirring on the output of a bright platinum electrode in a solution $1.33 \times 10^{-4}M$ $Fe_a^{+2} = Fe_a^{+3}$ and 1N HCl with H_2 gas continuously bubbled.

Where $E_{H_2}^o$ is the formal potential for the reaction, $H_2 = 2H^+ + 2e^-$ and equals O, E_{Fe}^o is 771 mV, f is the fraction of the electrode covered by the iron couple, and $1 - f$ the fraction covered by the hydrogen couple.

It was assumed that the reaction of iron or hydrogen with the platinum surface could be portrayed as a mass action equation

$$Fe^{+2} + Fe^{+3} + H_2 - Pt - H_2^{+2} \rightleftharpoons H_2 + 2H^+ + Fe^{+2} - Pt - Fe^{+3} \qquad [4]$$

from which the following equation was derived:

$$K_s = \frac{f}{(1-f)} \cdot \frac{[(H_2)(H^+)^2]^{1/2}}{[(Fe^{+2})(Fe^{+3})]^{1/2}} . \qquad [5]$$

Under standard conditions where $H_2 = 1$, $H^+ = 1$ and $Fe_a^{+2} = Fe_a^{+3}$ the measured Eh = 771 f. The activities of Fe_a^{+2} and Fe_a^{+3} were kept equal to each other and varied from 5×10^{-5} to 10^{-2} M in 1M HCl. Activity coefficients were estimated from mean activity values. The results of these measurements were used to compute K_s ($K_s = 1.17 \times 10^4$). The predicted Eh is shown in Fig. 5. The fit to the predicted equation is very good when the iron couple dominates, but is poor where the hydrogen couple dominates. It was assumed that a limiting current exists at low levels of the $Fe_a^{+2} - Fe_a^{+3}$ couple and this relationship is plotted in Fig. 6. The data also show that the relationship is influenced by stirring and whether a button or wire type electrode is used.

If the pH is increased to where iron precipitates it can be shown by Eq. [3 and 5] that above pH 4 the hydrogen couple will dominate. This was tested and found to be the case.

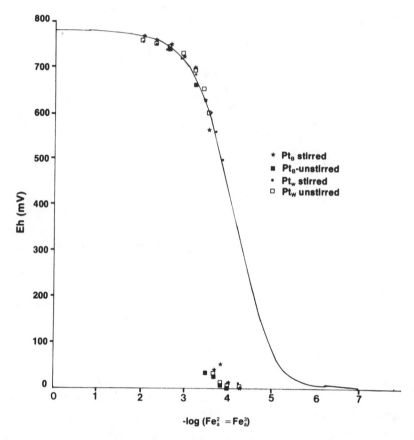

Fig. 5. Eh measured with various bright platinum electrodes as a function of ($Fe_a^{+2} = Fe_a^{+3}$) activities in 1N HCl with continuously bubbled H_2 gas. The curve was computed from Eq. [3] and [5] using a K_s value of 1.17×10^4.

A value of 11,700 was used for K_s and the Eh predicted by Eq. [3 and 5] calculated for pH values ranging from 4 to 7 with various ratios of Fe^{+3} to Fe^{+2}. It was assumed that a freshly precipitated $Fe(OH)_3$ was formed and $Fe_a^{+3} = 10^5(H^+)^3$. Observations indicated that at pH 3.8 $Fe(OH)_3$ was precipitated from a solution 10^{-2} in Fe^{+2} and Fe^{+3}. The Eh under N_2 gas was 480 mV and it was estimated that Fe^{+3} was 10^{-7} and Fe^{+2} was 10^{-2} and that these concentrations are low enough so that the $H_2 - H^+$ couple was predicted to be equally effective as the iron couple. The Eh under H_2 gas was -214 mV which would correspond approximately with pH 3.7 if the $H_2 = H^+$ couple controlled and the Pt electrode behaved as a hydrogen pH electrode. It was concluded from these observations that above pH 4 if H_2 gas were present the Pt electrode would not indicate the position of the $Fe^{+2} = Fe^{+3}$ couple. At pH levels below 4 the Pt electrode may give a reliable indication of the $Fe^{+2} - Fe^{+3}$ couple. An attempt to quantitatively

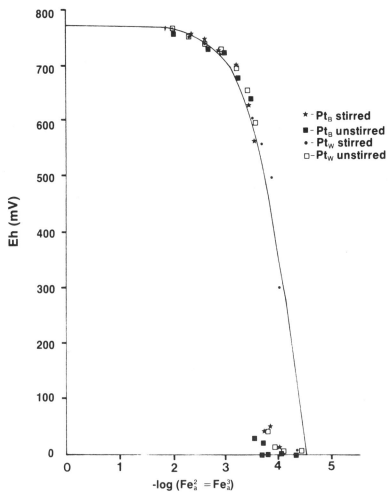

Fig. 6. Eh measured with various bright platinum electrodes as a function of $(Fe_a^{+2} = Fe_a^{+3})$ activities in 1N HCl with continuously bubbled H_2 gas. Curve computed from Eq. [3] and $f = 1(1 + Kd(c - 3 \times 10^{-5}))$ where K_s had a value of 1.22×10^4.

predict the position of the $Fe^{+2} - Fe^{+3}$ equilibria above pH = 4 with Eh-pH measurements would appear extremely unreliable.

Eh AND THE DEGREE OF OXIDATION OF SOILS

Easily reduced iron has been used to estimate the degree of soil oxidation (Coffin, 1963). A more general technique would be to titrate those components which are readily oxidized by a moderate oxidant. One purpose was to ascertain the relative oxidation state of soil and to relate this to Eh in an attempt to elucidate the utility of the Eh measurement.

Eh and pH Measurements

Eh measurements were made with button type bright platinum electrodes which have a surface area of approximately 0.4 cm². A calomel reference electrode was placed in saturated KCl and electrical contact made to the sample in a 30 cc vial by a salt bridge of 1 mm diameter filled with saturated KCl and 1.5% Agar-Agar. The output of the electrode was monitored through an Orion model 801 pH meter and continuously recorded. The pH was measured with the same system except that the glass electrode was substituted for the platinum electrode.

The Eh and pH values reported are those obtained 5 min after inserting the electrode. Usually there was a change in the Eh reading of 50 or more mv during the first 3 to 4 min after inserting the electrode followed by a gradual change of around 1 to 2 mv per hour (Fig. 7). Both pH and Eh measurements were made after the period of rapid change which in each case was observed on a strip chart recorder.

Estimate of Electron Acceptance of Soil

Five g of soil were placed in a 30 cc vial with a ground glass stopper and 5 ml of 0.197N Br_2 in glacial acetic acid added. The mixture was shaken, allowed to stand for 30 min, placed on a magnetic stirrer, and the excess Br_2 titrated with 0.100N As_2O_3 (4.941 g As_2O_3 plus 75 ml 1N NaOH neutralized to pH 7 with HCl and diluted to 1 liter). The end point of the titration was determined by continuously recording the output of a platinum electrode to find the point of maximum Eh change per unit of As_2O_3 added as shown in Fig. 8.

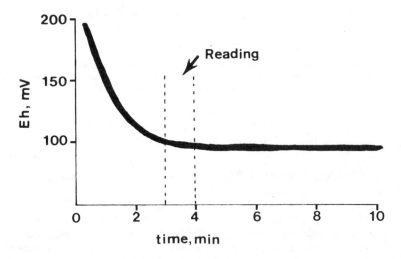

Fig. 7. Average drift observed for freshly cleaned Pt electrodes inserted into soil water mixtures.

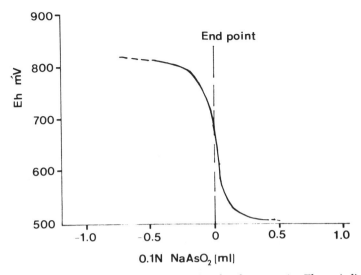

Fig. 8. Titration of excess Br₂ in glacial acetic acid with sodium arsenite. Eh was indicated by the output of a button type platinum electrode.

Easily Reducible Iron and Manganese

One g of soil was placed in a 50 cc centrifuge tube and 1 g sodium hydrosulfide and 20 ml of 0.2M sodium citrate at pH 4.75 added (Coffin, 1963). The mixture was heated for 30 min at 50C, filtered, and 5 ml placed in a 50 ml digestion tube. Five ml of concentrated HNO_3 was added and the mixture heated until dry. The residue was dissolved in 5 ml concentrated HCl and diluted to 50 ml. Iron was determined by the orthophenanthroline (Golterman, 1969) method and manganese by atomic absorption.

BIOLOGICAL ESTIMATES OF ELECTRON ACCEPTANCE CAPACITY

Various amounts of glucose were added to 5 g samples of A, B, and C horizons of Menfro and Mexico soils placed in 30 cc vials. The 1 soil to 2 water mixtures were equilibrated for 6 to 15 days and the Eh, pH, and electron acceptance measured as described.

The Eh values of the Ap horizon of the Mexico soil to which glucose was added are plotted as a function of time and shown in Fig. 9. The pattern for B and C horizons was similar with the following exceptions. A minimum Eh in the Ap horizon sample was reached after 1 day, the C horizon at 2 days and the B22 horizon at 4 days. After 7 days the Eh was relatively constant and from 100 to 400 mv higher than the minimum in all horizons. Measurements of the amount of reduction were made at 6 and 15 days after addition of glucose.

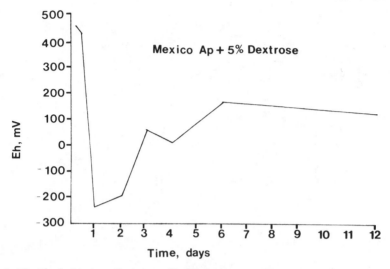

Fig. 9. The Eh of a Mexico soil submerged in 5% dextrose at various times after addition.

The titration of excess added Br_2 was carried out as described and a titration curve is shown in Fig. 8. The end point is very distinct with an Eh drop of 250 mv resulting from the addition of 0.1 ml of 0.1N As_2O_3. The end point for soil titrations could be determined to the nearest 0.05 ml of titer. The reason for adding excess Br_2 instead of directly titrating the soil with Br_2 was that this was too slow a reaction for a practical titration.

The idea of the Br_2 titration was to select an oxidizing agent which would oxidize Fe^{+2} to Fe^{+3} or Mn^{+2} to Mn^{+4} without oxidizing more resistant organic components. When Br_2 was added to glucose and the excess titrated with As_2O_3 no oxidation of glucose was indicated. It has been shown that Br_2 does not oxidize protein, saturated fats, or carbohydrates. However, easily oxidizable groups such as unsaturated fats will be oxidized by bromine. The effect of amount of glucose added on the electron acceptance capacity of Menfro and Mexico is shown in Fig. 10 and 11. The electron acceptance was not greatly increased by increasing the amount of glucose beyond 2,400 mg glucose/100 g soil in either the Menfro or Mexico soils. In both soils the electron acceptance capacity of the A11 or Ap horizons was higher than A2 or B22 horizons which was higher than B33 or C horizons.

The Eh was measured prior to the bromine titration in each case. The relationships between Eh and electron acceptance capacity for Menfro and Mexico soils are given in Fig. 12. For each 31.5 mV increase in Eh the electron acceptance capacity decreased 1 me⁻/100 g soil. Seventy-two percent of the variation in Eh was accounted for by changes in electron acceptance capacity. Due to "perched potentials", slow reactions, and interactions of the electrode surface with soil components the Eh measurement is of dubious quantitative value. However, data presented in Fig. 12 show that the Eh measurement is related to the general oxidation level of the soil. It was concluded that using Eh in diagrams to predict the relative

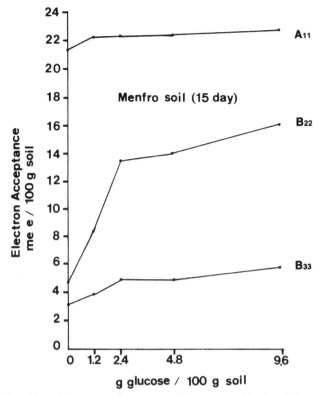

Fig. 11. The effect of glucose on the amount of Br_2 consumed by Ap, B22, and Cl horizon samples of a Mexico soil.

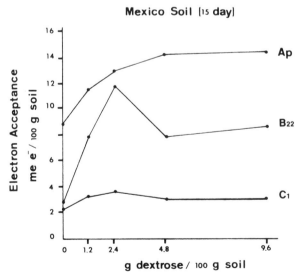

Fig. 10. The effect of glucose on the amount of Br_2 consumed by A11, B22, and B33 horizon samples of a Menfro soil.

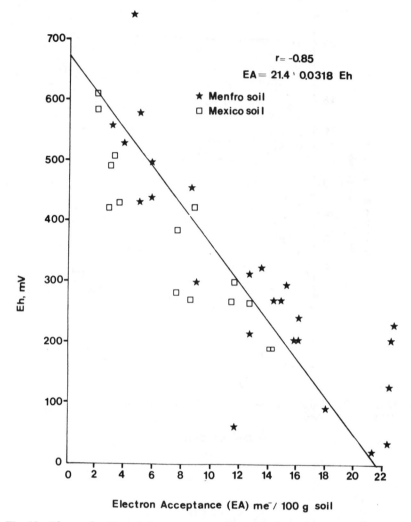

Fig. 12. Eh as a function of electron acceptance for Menfro and Mexico soils submerged under water or dextrose solutions for various times.

stability of various minerals and the forces affecting their transformations is of value. It would be predicted that comparisons of computed levels of constituents such as iron or manganese in soils based on Eh and pH measurements would be highly unreliable unless large amounts of the ions in question were in solution.

Eh-Ph MEASUREMENTS IN SOILS IN PLACE

Platinum electrodes constructed as shown in Fig. 13 were installed in a Mexico soil (Hess and Blanchar, 1977). The electrode consisted of a plastic water pipe 1.3 cm in diameter and insulated copper wires with 22

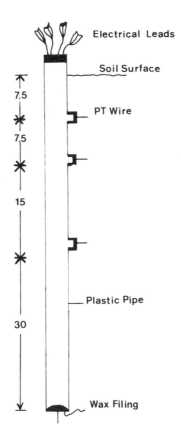

Electrical Leads

Soil Surface

7.5

PT Wire

7.5

15

30

Plastic Pipe

Wax Filing

Fig. 13. Platinum electrode arrangement constructed by Dr. R. E. Hess and buried in the Mexico soil.

gauge platinum wire soldered on one end and a female jack on the other. The wires were placed in the tube so that 2 cm of the platinum would be exposed, and the tube was filled with paraffin to insulate the wires inside the tube against water. The top end of the electrode had one female jack for each platinum wire. Four duplicate electrodes were placed at depths of 7.5, 15, 30, and 60 cm below the soil surface. A 7.5 diameter hole was made with a core sampler. The electrode was placed in the soil so that the platinum wires went gently into the wall of the hole without upsetting the soil. Soil from the core was replaced in the hole by breaking it up according to depth and replacing it with occasional tamping. A second hole 40 cm deep was made and a 40 by 2 cm plastic pipe with its end plugged was inserted. This pipe was removed and the hole moistened with $0.01M$ $CaCl_2$ and the calomel electrode inserted when Eh measurements were made.

Eh and pH measurements were made using a portable pH meter, platinum electrodes, a sleeve type calomel reference electrode, and a glass electrode. Samples for pH measurements were obtained at the proper depths for the measurement using a sampling tube. The soil sample was placed in a plastic beaker, an equivalent amount of water added, and the pH measured.

A similar study was carried out on the Menfro soil where in addition to Eh and pH the soil moisture content and composition of soil air in terms of N_2, CO_2, and O_2 were measured (Feagley, 1979).

The Eh at the 15 cm depth indicated that the Mexico soil was oxidized through most of the year (Fig. 14). Periods of prolonged low Eh were observed at both the 30 cm and 60 cm depth in the Mexico soil. The periods of low Eh were longer and the Eh values lower at the 60 cm depth than those at the 30 cm depth.

The Eh at 60 cm gradually increased through June, July, and August as the soil became dry. In early September, 3.84 inches of rain fell in 3 days and the Eh dropped. Feagley (1979) measured Eh and percent moisture on the Menfro soil. Prior to any addition of water the moisture content of the 30 cm layer was 38% and the 60 cm layer 48%. After a period of excessive watering the Eh at the 60 cm depth dropped from 600 mV to −100 mV and the moisture content remained constant near 48%, however at the 30 cm depth volumetric moisture increased from 38 to 46%. It appeared that Eh decreased after the 60 cm layer was isolated enough to reduce air movement to it.

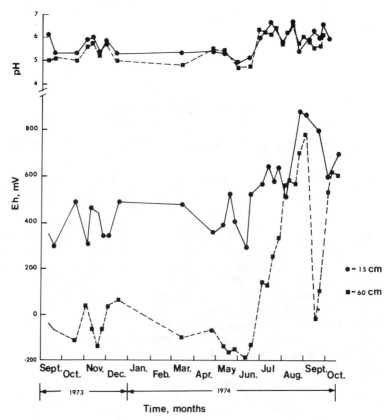

Fig. 14. Eh and pH measurements at depths of 15 and 60 cm in a Mexico soil during 1973 and 1974.

Stability diagrams for iron and manganese were constructed using a free energy of formation of -166 kcal/mole for freshly precipitated $Fe(OH)_3$ (Garrels and Christ, 1965) and an E^0 (880 mV) value for MnO_2 being reduced to Mn^{+2} (Ponnamperuma et al., 1969). The reaction considered for $Fe(OH)_3$ and shown in Fig. 15 may be written as:

$$Fe^{+2} + 3H_2O \rightleftharpoons Fe(OH)_3 + 3H + e^-, \qquad [6]$$

while the reaction for Mn^{+2} in Fig. 16 was:

$$Mn^{+2} + 2H_2O \rightleftharpoons MnO_2 + 4H^+ + 2e^-. \qquad [7]$$

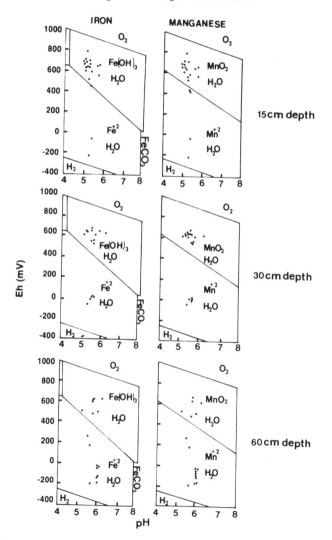

Fig. 15. Yearly variations in Eh and pH in various depths of a Menfro soil as related to iron and manganese stability (Feagley, 1979).

Eh and pH points measured through the summer months for Menfro and Mexico soils were plotted on the iron and manganese stability diagrams. The Fe^{+2} and Mn^{+2} level depicted was $10^{-6}M$. Only on a few occasions and for periods of short duration did the Eh drop low enough to indicate that either iron or manganese compounds in the 15 cm depth of either soil would dissolve (Fig. 14, 15, 16). In both soils at the 30 cm depth the Eh was low enough and persisted for several weeks so that increased dissolution of iron and manganese oxides would be expected. At the 60 cm depth there were prolonged periods of low enough Eh to indicate that iron and manganese oxides would not be stable (Fig. 14, 15, 16).

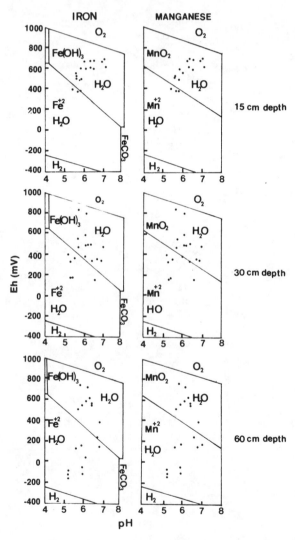

Fig. 16. Yearly variations in Eh and pH in various depths of a Mexico soil as related to iron and manganese stability (Hess and Blanchar, 1977).

Feagley (1979) measured iron and manganese concentrations in dilute calcium chloride extracts of samples of Menfro soil after a period of excessive water application. He concluded that changes in manganese concentration in these extracts were more closely related to changes in CO_2 content of the soil air than to changes in Eh. It may be that due to the lag period involved between dissolution or precipitation and changes in oxidation state a quantitative relationship between a given Eh measurement and manganese solubility did not occur.

The extent and duration of Eh changes in soils due to variations in the degree of aeration may be related to the capacity of soil components to give or take electrons. Mattson (1943) measured the Eh of mixtures of oxygen saturated and unsaturated humus and concluded that the reductive capacity of humus at a different pH can be determined by oxygen consumption. It has generally been assumed that estimates of easily reducible iron and manganese in mineral soils provide an estimate of Eh buffering. Allison and Scarseth (1942) pointed out that iron extracted from soils reduced by biological activity under anaerobic conditions was of the same order of magnitude as that extracted with sulfides and oxalic acid. Electron acceptance capacity and easily reducible iron and manganese are compared in Table 1. In the surface horizon of the Mexico soil the values were of the same magnitude, however, electron acceptance capacity values were lower than chemically reducible iron and manganese in all subsoil samples. This may be due to incomplete biological reduction in the subsoil samples and to inadequate nutrition for the biological processes.

The validity and interpretation of the Eh measurement when electrodes are implanted in soils may be considered from several viewpoints. Both Hess and Blanchar (1977) and Feagley (1979) observed that during periods of rapid change, Eh measurements from several electrodes placed at the same depth were extremely variable. Feagley (1979) determined the value and standard deviation for five platinum electrodes placed at a 120 cm depth in the Menfro soil and read weekly from 15 June 1978 to 1 Sept. 1978. This depth and period was chosen because the moisture content was constant. He reported the Eh values for the five electrodes to be 660 ± 50, 560 ± 400, 600 ± 70, 660 ± 40, and 670 ± 40 mV. He assumed the second electrode was malfunctioning and concluded that a properly functioning platinum electrode would have an expected reproducibility of between 50 ± 10 mV.

Table 1. Electron acceptance capacity and easily reducible iron and manganese in Menfro and Mexico soils.

	Biologically reduced	Chemically reduced	
	EAC	Fe	Mn
	me e⁻/100 g	—— mM/100 g ——	
Menfro A11	22.5	9.8	1.5
Menfro B22	16.1	25.3	1.0
Menfro B33	5.0	23.0	1.2
Mexico Ap	12.8	12.8	1.8
Mexico B22	8.3	16.3	1.2
Mexico C1	2.8	10.9	0.2

If the measured values of Eh are considered to be reproducible what are the chemical forms responsible for the potentials measured by the electrode? Stumm and Morgan (1970) explained that an inadequate exchange current of about 10^{-7} ampere would result from a solution 10^{-6}M in both Fe_a^{+3} and Fe_2^{+2} and the Eh measurement of this couple would no longer be precise. The exact value for the concentration of $Fe_a^{+2} - Fe_a^{+3}$ at which limited detection occurs would depend upon such factors as electrode size and other competitive perched oxidation-reduction reactions. These observations are consistent with those observed in our studies where a limit of 3×10^{-5}M Fe_a^{+2} and Fe_a^{+3} in the presence of 1 atmosphere H_2 gas pressure was a limiting current.

A reproducible Eh has been observed in systems at pH = 7 which contain precipitated $Fe(OH)_3$. The reaction is described in Eq. [6] and can be used to predict iron equilibria if Fe_a^{+2} in solution is greater than 10^{-6}M (Fig. 15 and 16). Doyle (1968) believed that this reaction yielded a reproducible Eh due to the formation of a ferric hydroxide coating on the platinum surface which catalyzed the electron transfer. Feagley (1979) observed that when Mn^{+2} was reacted with O_2 in the presence of dilute NaOH a coating of birnesite ($MnNaO(OH)_3$) formed on the platinum surface. It is believed that the platinum electrode responds poorly to Mn^{+2}, but responds well to the $Fe^{+2} - Fe^{+3}$ couple and that it indicates a mixed potential depending on the degree of interaction of Mn^{+2} with the $Fe^{+2} - Fe^{+3}$ couple (Stumm and Morgan, 1970).

Bohn (1968, 1969) concluded that Eh measured with a platinum electrode is a rough measure of the system's oxidation state. It was concluded by Bohn (1968) that at high Eh it is a mixed potential involving the O_2-H_2O couple and at low Eh involving the H_2-H_2O couple, but both are mixed potentials which are not quantitatively related to the distribution of oxidized and reduced species of any given reaction. This interpretation appears valid from studies reported here.

Placing a Menfro surface soil under water for 20 days increased iron from 2 to 18 μg Fe/g of soil and manganese from 1 to 22 μg Mn/g of soil (Hess and Blanchar, 1977). When dextrose was added the Eh became even more negative and the iron content reached 458 μg Fe/g and manganese 159 μg Mg/g. After the samples were dried Eh increased and both iron and manganese decreased. Feagley (1979) found higher levels of soluble iron and manganese in the Menfro soil in the field when it was watered excessively. However these iron and manganese values were not quantitatively related to those predicted from Eh and pH measurements.

Blanchar and Scrivner (1972) found the ion product pFe + 3pOH = 41.1 in the Menfro soil to be close to that predicted for goethite from Langmuir's (1969) values. Kao and Blanchar (1973) reported values of pFe + 3pOH = 37.6 for surface horizons, and about 39 for C horizons of the Mexico soil. These measurements were made in dilute HCl solution with pH levels in the 2.5 to 4.5 range in both cases. The addition of stronger acid in the case of the Mexico soil reduced the pH and increased the value of pFe = 3pOH to 40.6 near that of the Menfro soil and of goethite. These changes in pFe + 3pOH are small compared to those predicted from observed changes in soluble iron, pH, and Eh for Menfro and Mexico soils in place. It is postulated that failure to predict iron or manganese content of soil solution in the field is not caused by not knowing what controls the solubility, but inability to measure the appropriate Eh, pH, ionic activities, and ionic strength required for the computation.

CONCLUSIONS

1. If the ion couples to be measured by the Eh electrode are present in high enough concentrations and are electro-active, quantitative use of Eh and pH to predict the position of equilibria involving that couple are possible. Also if couples present in lower concentrations are known to react rapidly and quantitatively with the dominant current producing couple, valid Eh and pH measurements are possible.

2. Mineral equilibria can best be estimated if PO_2 and pH_2 pressures in the mixtures are reduced as far as possible. With low concentrations of iron or manganese the presence of oxygen will give a higher mixed potential of little quantitative value and the presence of hydrogen a lower value which cannot be interpreted. If the objective of the measurement is to attempt quantitative prediction of the equilibrium position of mineral suites the system should be purged with N_2.

3. Eh and pH measurements indicate changes which alter solution composition by several million fold in soil systems in place. Although their uses for quantitative calculation of solution composition is limited they are extremely valuable for predicting the direction of change and the compounds which probably control the solubility.

LITERATURE CITED

1. Allison, L. E., and G. d. Searseth. 1942. A biological reduction method for removal of iron oxides from soils and colloidal clays. J. Am. Soc. Agron. 34:616–623.

2. Bass Becking, L. G. M., I. R. Kaplan, and D. Moore. 1960. Limits of the natural environment in terms of pH and oxidation-reduction potential. J. Geol. 68:243–284.

3. Blanchar, R. W., and C. L. Scrivner. 1972. Aluminum and iron ion products in acid extracts of samples from various depths in a Menfro soil. Soil Sci. Soc. Am. Proc. 36:897–901.

4. Bohn, H. L. 1968. Electromotive force of inert electrodes in soil suspension. Soil Sci. Soc. Am. Proc. 32:211–215.

5. ————. 1969. The emf of platinum electrodes in dilute solutions and its relationship to soil pH. Soil Sci. Soc. Am. Proc. 33:639–640.

6. ————. 1970. Comparison of measured and theoretical Mn^{+2} concentrations in soil suspension. Soil Sci. Soc. Am. Proc. 34:195–197.

7. Coffin, D. E. 1963. A method for the determination of free iron in soils and clays. Can. J. Soil Sci. 43:7–17.

8. Collins, J. F., and S. W. Buol. 1970. Effects of fluctuation in the Eh-pH environment on iron and/or manganese equilibria. Soil Sci. 110:111–118.

9. Doyle, R. W. 1968. The origin of the ferrous ion-ferric oxide N potential in environments containing dissolved ferrous iron. Am. J. Sci. 226:840–859.

10. Eisenman, George. 1969. Theory of membrane electrode potentials: an examination of the parameters determining the selectivity of solid and liquid ion exchanges and of neutral ion-sequestering molecules. p. 1–56. In R. A. Durst (ed.) Ion selective electrodes. Nat. Bur. Stand. Publ. 314.

11. Feagley, S. E. 1979. Rates of manganese oxidation in a Menfro soil. Ph.D. Thesis, Univ. of Missouri-Columbia. In registration.

12. Garrels, R. M., and C. L. Christ. 1965. Solutions, minerals, and equilibria. Harper and Row, New York.

13. Golterman, H. L. 1969. Methods of chemical analysis of fresh waters. IBP Handbook No. 8.

14. Gotoh, S., and W. H. Patrick, Jr. 1974. Transformations of iron in a waterlogged soil as influenced by redox potential and pH. Soil Sci. Soc. Am. Proc. 38:66–71.

15. Hess, R. E., and R. W. Blanchar. 1977. Arsenic determination and arsenic, lead, and copper content of Missouri soils. Mo. Agric. Exp. Stn. Bull. 1020. p. 1–46.

16. Kao, C. W., and R. W. Blanchar. 1973. Distribution and chemistry of phosphorus in an Albaqualf soil after 82 years of phosphate fertilization. J. Environ. Qual. 2:237–240.

17. Langmuir, D. 1969. Geochemistry of iron in a coastal-plain ground water of the Camden, New Jersey area. U.S. Geol. Surv. Proc. 650-C:C224–C235.

18. ————, and D. O. Whittemore. 1971. Variations in the stability of precipitated ferric oxyhydroxides. Adv. Chem. 106:209–234.

19. Marshall, C. E. 1977. Physical chemistry and mineralogy of soils. II. Soil in place. John Wiley & Sons, New York.

20. Mattson, Sante. 1943. The oxidation-reduction condition in vegetation litter and humus: I. Reduction capacity, oximetric titration and liming toxicity. Ann. Agric. Coll. Sweden 2:135–144.

21. Morris, J. C., and W. Stumm. 1967. Redox equilibria and measurements of potentials in the aquatic environment. Adv. Chem. 67:270–285.

22. Pasricha, N. S., and F. N. Ponnamperuma. 1976. Influence of salt and alkali on ionic equilibria in submerged soils. Soil Sci. Soc. Am. J. 40:374–376.

23. Ponnamperuma, F. N., E. M. Tianco, and T. Loy. 1967. Redox equilibria in flooded soils: I. The iron hydroxide systems. Soil Sci. 103:374–382.

24. ————, T. A. Loy, and E. M.Tianco. 1969. Redox equilibrium in flooded soils. II. The manganese oxide systems. Soil Sci. 108:48–57.

25. Sato, Motoaki. 1960. Oxidation of sulfide ore bodies. 1. Geochemical environments in terms of Eh and pH. Econ. Geol. 55:928–961.

26. Stumm, W., and J. J. Morgan. 1970. Aquatic chemistry; an introduction emphasizing chemical equilibria in natural waters. John Wiley and Sons, Inc., New York.

27. Whitfield, M. 1969. Eh as a operational parameter in estuarine studies. Limnol. Oceanogr. 14:547–558.

28. ————. 1974. Thermodynamic limitations on the use of the platinum electrode in Eh measurements. Limnol. Oceanogr. 19:857–865.

29. Yamane, I. 1973. Eh-pH diagrams of manganese systems in relation to flooded soils. Rep. Inst. Agric. Res. Tohoku Univ. 24:1–15.

QUESTIONS AND ANSWERS

Q. How did you take the drift that would come over time into account?

A. We did not. We presumed that the response time of the platinum electrode was rapid and by inserting it we altered the system, and so we took that reading after the rapid potential change ceased, but slow change had not, recognizing that what we were dealing with mixed potentials in a system not at equilibrium.

Q. What was the pH at which you ran those tests with equal amounts of iron?

A. The pH of the 1 molar hydrochloric acid system was zero.

Q. You made the assumption in your iron system that you were dealing with iron in solution. What about structural iron in your clay minerals?

A. We were dealing with the iron in solution and did not consider the source in terms of the Eh measurements that we've made. There have been observations that iron coatings do form on the platinum electrode and it acts like an iron hydroxide electrode; but we did not consider this in the interpretations of our data.

Q. Do you think then that the Eh measurements of iron in solution would be the same as Eh measurement for structural iron in clay minerals?

A. No, I don't think they would be the same. I presume that the structural iron in clay minerals indicated by solubility-type measurements would be much more insoluble.

Q. What's going to be the impact on your field measurements when the electrodes are not uniformly or even completely wet?

A. We have observed in drying cycle a total loss of contact and the meter behaves as an open circuit. If electrodes were partially wetted, I would suspect that we would have reduced exchange currents which would be indicated by instability and drift in our readings which would add uncertainty about concentrations of the couples of interest in solution.

Q. When you get a cycle of iron reduction and re-oxidation once a year, would you say that the iron oxides in the subsoil are being formed and redesorbed and reformed once a year?

A. I think that's correct. I presume, in the Mexico soil, the one most poorly aerated, that we have a more reactive iron hydroxide than we have in our better aerated Menfro soil. The solubility product of the iron hydroxides in dilute acid for the Mexico soil indicated values close to that for amorphous iron hydroxide, about a pK of 38.5, and those for the Menfro soil were around 40 to 41 indicating goethite.

Q. What was the variability of the Eh electrodes at any one depth in the field?

A. We deduced that in this manner. We had 12 electrodes buried 120 cm in the Menfro soil and our observations of volumetric moisture content indicated that it did not change. The standard deviation was ± 50 mV.

Q. At the end of your experiment, did you pull out your electrodes and check them?

A. No, we haven't yet.

Q. What is your experience as far as the memory effect is concerned?

A. Well, we find it very difficult to tell under field conditions because we don't know what the Eh should be.

Q. In any of your work did you take an electrode out and check the equipment system?

A. Yes, and there is a memory in the system. If we have been in a re-
    duced system the values will be low until we physically abrade the
    platinum surface.

Q. Did you correct for the chloride complexing of the ferrous species
    going from the high concentrations to the low concentrations? Be-
    cause perhaps the relative activities would change.

A. I looked at the values that were available in Sillen and Martell and
    they indicated that there would not be stable chloride complexes in
    that system. I haven't looked for more recent values. I did not correct
    for chloride complexes.

CHAPTER 7

Metal-Organic Matter Interactions in Soil

P. R. BLOOM[1]

The binding of metallic cations by soil organic matter has a profound influence on the physical and chemical properties of soils. The cation exchange capacity (CEC) of organic soils is largely due to organic matter and its contribution to the CEC of mineral soils may be > 200 meq/100 g of organic matter (Helling et al., 1964). Humic substances are important both in the retention of micronutrients in the soil (Cheshire et al., 1977) and in their transport in soil solutions (Geering et al., 1969). The binding of iron (Fe) and aluminum (Al) by humic substances is an important factor in spodosol formation (Martin and Reeve, 1958) and in the stabilization of soil aggregates (Giovannini and Sequi, 1976). The exchange of Al from organic matter binding sites in acid soils is important in controlling soil solution Al levels and neutral salt extractable Al (Bloom et al., 1979a; 1979b).

---

[1] Dep. of Soil Science, Univ. of Minnesota, St. Paul, MN 55108.

Soil organic matter is largely composed of polymeric weak acids. In the pH range found in soils, metal ions are bound largely by carboxyl groups. Some of the more strongly bound ions may also bind to phenolic groups. The exact nature of the binding sites and the nature of interaction of cations with these sites has been the subject of recent controversy. This paper will critically review the research of organic matter interactions with cations, with an emphasis on the more recent work.

COMPOSITION OF HUMIC POLYMERS

Humic acid and fulvic acid are the organic matter polymeric fractions which have been most studied. Humic and fulvic acids are extracted from a soil with alkaline extracting agents, usually NaOH or $Na_4P_2O_7$. Humic acid is precipitated with mineral acid and purified by dialysis against acid or water. A treatment with HF may also be used to assure a very low ash content [see Posner (1966), for a comparative study of 16 different methods of preparing humic acid]. Fulvic acid is the fraction that is acid soluble.

Methods for extracting humic and fulvic acids lead to recovery of only a fraction of soil organic matter, usually about 10 to 70% (Posner, 1964; Flaig et al., 1975).

Humic and fulvic acids appear to be quite similar in composition but with different molecular weights. Research on the composition of these polymers has been reviewed by Flaig et al. (1975), Schnitzer (1978b) and Hayes and Swift (1978). The more acidic nature of fulvic acid compared to humic acid is shown by its greater content of carboxyl groups (Table 1).

Humic and fulvic acid polymers have considerable aromatic character and have been thought to be somewhat similar in structure to lignin, a natural polyphenol (Flaig et al., 1975). Recent nuclear magnetic resonance (NMR) evidence, however, suggests that alkyl chains make important contributions to humic structure. In a study of a soil humic acid, Wilson et al. (1978) found that only about 65% of the carbon occurred in aromatic or carboxyl units.

Humic acid polymers are colloidal in size with average molecular weights of about 5×10^4 (Flaig et al., 1975, Hayes and Swift, 1978). Flaig et al. (1975) concluded that humic acid particles are spherical with diameters of 80 to 100 Å. This is suggestive of a highly cross-linked structure (Bloom and McBride, 1979). Hayes and Swift (1978) concluded that a random coil model is more appropriate, while Schnitzer (1978a) suggests that humic acids behave more like linear polyelectrolytes.

Less is known about the molecular configuration of fulvic acids. The molecular weights of fulvic acids are, however, much less than that of humic acids (Schnitzer, 1978b).

Humic and fulvic acids have nitrogen (N)-containing groups in addition to the oxygen-containing functional groups listed in Table 1. The N-containing groups are of interest because of the high affinity of some transition metal ions [e.g. nickel (Ni) and copper (Cu)] for N-containing ligands. Hydrolyzates of soil organic matter contain about 50% amino N (Flaig et al., 1975). Much of this N is likely involved in peptide bonds. The remainder of the N is not well defined, but some may be heterocyclic N.

Table 1. Analysis of typical humic and fulvic acids (Schnitzer, 1978a).

Element	HA	FA
%		
C	56.2	45.7
H	4.7	5.4
N	3.2	2.1
S	0.8	1.9
O	35.5	44.8
Total	100.4	99.7
	Functional group	
	meq/g	
Total acidity	6.7	10.3
CO_2H	3.6	8.2
Phenolic OH	3.9	3.0
Alcoholic OH	2.6	6.1
Quinonoid C = 0 Ketonic C = 0	2.9	2.7
OCH_3	0.6	0.8

Humic and fulvic acids contain a small quantity of sulfur (S) (Table 1). Much of this is reduced S (Anderson, 1975) which could be important in binding small quantities of "soft" ions like cadmium (Cd^{2+}) and mercury (Hg^{2+}).

Binding by groups other than COOH and phenolic OH groups contributes only a minor fraction to the CEC of soil organic matter. Blocking OH and COOH groups by exhaustive methylation with diazomethane reduced the CEC of soil organic matter and peat (as measured by barium acetate) to near zero (Broadbent and Bradford, 1952). Another study showed that methylation of humic acid reduced the retention of Cu^{2+}, after 8-1N HCl washes, by 40% (Davies et al., 1979); however, the Cu^{2+} retained after HCl washing was only 0.1% of that retained at saturation. Thus, other functional groups may participate in the binding of small quantities of some metals.

Attempts to block either phenolic OH or COOH groups selectively have led to conflicting conclusions. Gillam (1940) found no reduction in CEC when OH groups were blocked with dimethylsulfate while Broadbent and Bradford (1952) found considerable reduction in CEC using the same reagent. From their selective blocking experiments, Schnitzer and Skinner (1965) concluded that COOH and phenolic OH cooperate in the binding of Fe^{3+} and Cu^{2+}. They suggested that soil organic matter contains salicylic acid-type groups (Fig. 1). Interpretation of these results is difficult because of the possibility of side reactions in a functionally heterogeneous system especially under the drastic reaction conditions used (e.g. refluxing 24 hr in acetic anhydride plus H_2SO_4).

Infrared analysis of the product formed by refluxing humic acid with acetic anhydride suggested that a significant quantity of COOH groups are close enough together to form 5 to 7 member ring anhydrides, (Wood et al., 1961; Wagner and Stevenson, 1965; Stevenson and Goh, 1974). The existence of phthalic acid-type functional groups (Fig. 1) in humic acid has been suggested by Wood et al. (1961).

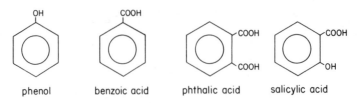

Fig. 1. Model weak acid sites in soil organic matter.

SPECTROSCOPY

Infrared

Infrared absorption spectroscopy (IR) is a useful tool to study metal ion binding in organic matter. Infrared absorption is due to interaction of IR photons with molecular vibrations. Bond types or molecular structures absorb at characteristic frequencies. A prominent feature of the IR spectra of fulvic and humic acids (Fig. 2) is an absorption peak at about 1700 cm^{-1}. This absorption has been assigned to C = O stretching (Stevenson and Goh, 1971; Tan, 1977). The exchange of COOH protons with metal ions results in the disappearance of the absorption band at 1700 cm^{-1} and the appearance of new bands at about 1600 cm^{-1} and 1380 cm^{-1} (Banerjee and Mukherjee, 1972; McCarthy et al., 1975; Vinkler et al., 1976). The latter two bonds are due to symmetrical and antisymmetrical COO$^-$ stretching modes respectively (Vinkler et al., 1976).

The antisymmetric stretching of COO$^-$ is somewhat influenced by the humate cation. Vinkler et al. (1976) studied the influence of alkali and alkaline earth metals and Mn^{2+} (manganese), Fe^{2+}, Zn^{2+} (zinc), Cu^{2+}, Cr^{3+} (chromium), and Al^{3+} on the antisymmetric COO$^-$ absorption of dried humates. They reported absorption bands (± 5 cm^{-1}), to range from 1585 cm^{-1} for Ca^{2+} to 1625 cm^{-1} for Al^{3+}. A higher frequency indicates greater covalency. The bonding, however, is largely ionic. Under natural conditions cation hydration water may be an important factor in the binding of cations and covalency observed by Vinkler et al. (1976) may be less significant.

It is possible to study humates in the hydrated state either by replacing the water with D$_2$O or by the use of Fourier transform IR spectroscopy. Deuterium oxide does not absorb in the 1600 cm^{-1} region and Fourier transform allows for the subtraction of the water spectrum. Two limited studies with humic acid (McCarthy and Mark, 1975; McCarthy et al., 1975) have demonstrated the potential of these techniques.

Electron Spin Resonance and Nuclear Magnetic Resonance

Electron spin resonance spectroscopy (ESR), also called electron paramagnetic resonance (EPR), has been used to study metal ion binding in humates. Nuclear magnetic resonance (NMR), while used extensively to determine the structure of organic matter, has received little attention in the study of metal ion binding.

Na Humate

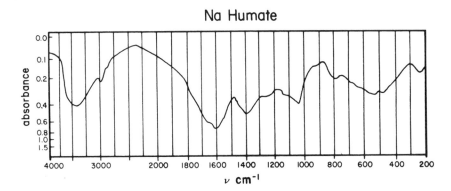

Humic Acid

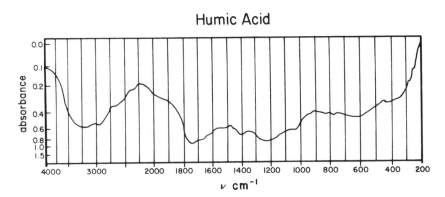

Fig. 2. Infrared absorbance of humic acid and sodium humate (Vinkler et al., 1976).

Electron spin resonance results from the absorption of microwave energy by unpaired electrons in an applied magnetic field. Electrons with spins parallel to the magnetic field are at a lower potential energy than antiparallel electrons. Absorption of a microwave photon results in a parallel to antiparallel transition. The microwave frequency at which resonance takes place is a function of the applied magnetic field and the interaction of the electron spin with the magnetic vector of an atomic nucleus. The spectra of ions show hyperfine splitting due to the quantization of nuclear spins.

Since ESR spectrometers vary the magnetic field at a fixed microwave frequency, the units of spectral data are in Gauss, (Fig. 3 and Fig. 4). Most spectrometers record the first derivative of the absorption spectrum (Fig. 3 and Fig. 4).

Most ESR studies on organic matter have involved the free radical signal (Fig. 3c). Humates have been suggested to contain semiquinone free radicals which have an ESR absorption at $g = 2.003$ (Schnitzer, 1978a; McBride, 1978; Wilson and Weber, 1977).

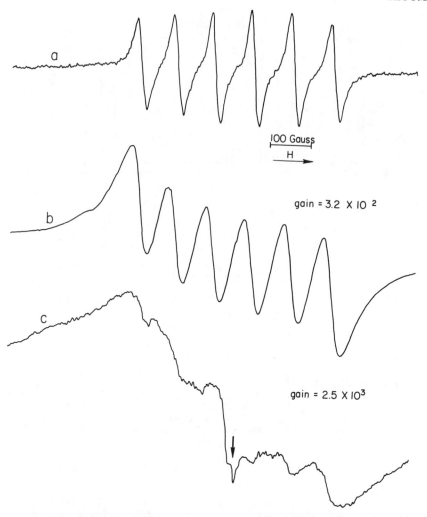

Fig. 3. Room temperature ESR spectra of a) $1 \times 10^{-4}M$ $MnCl_2$; b) 0.10 equivalent fraction of Mn^{2+} on wet H^+-peat; c) 0.10 equivalent fraction of Mn^{2+} on H^+-peat, dried at 110 C. Vertical arrow indicates the free electron signal in peat (g = 2.00) (Bloom and McBride, 1979).

Transition metal ions which have unpaired d electrons can be studied by ESR. Studies of Mn^{2+} ESR spectra in acid washed peat, (Bloom and McBride, 1979) and humic acid (McBride, 1978) suggest that Mn^{2+} does not form a chelation complex with these humic materials at moderately low pH. The spectrum of Mn^{2+} in wet H-peat, (Fig. 3b) is quite similar to the spectrum of Mn^{2+} in solution (Fig. 3a); however, the peat spectrum is broadened compared to the solution spectrum. The line widths in solution are 23 Gauss compared to 54 Gauss in the peat. Solution like spectra with broadened line widths have also been observed for Mn^{2+} in pectin, a natural polycarboxylic acid, in Amberlite IRC-50, a polycarboxyl resin (Lakatos et al., 1977b), and in a wet smectite (McBride et al., 1975). The

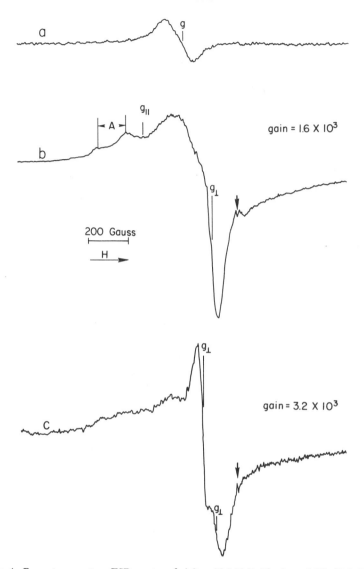

Fig. 4. Room temperature ESR spectra of a) $2 \times 10^{-3} M$ $CuCl_2$, (g = 2.18); b) 0.15 equivalent fraction of Cu^{2+} on H^+ peat, wet (g ‖ = 2.32, g ⊥ = 2.09); c) 0.15 equivalent fraction
of Cu^{2+} on H^+-peat, dried at 110 C, g ⊥ = 2.07 (and 2.11). Arrow indicates the free electron signal in peat (g = 2.00 (Bloom and McBride, 1979).

broadened solution-like spectrum suggests that Mn^{2+} is fully hydrated on
the exchange sites but that its rate of tumbling is restricted by the interaction with binding sites. Nuclear magnetic resonance spectroscopy of the
water protons in wet Mn^{2+} fulvate suggests that Mn^{2+} bound to fulvic acid
is also fully hydrated (Gamble et al., 1976).

The complex formed between humates and Mn^{2+} can be described as
an outer sphere complex where the ligand does not interact directly with

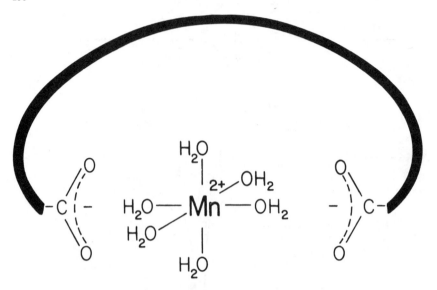

Fig. 5. Possible binding site for Mn^{2+} in soil organic matter.

the Mn^{2+} electrons (Fig. 5). This is not a chelation complex since a chelation complex is an inner sphere complex where functional groups donate electron pairs to a metal ion. Drying broadens the spectrum (Fig. 3c) due to spin-spin interaction between Mn^{2+} ions resulting from the reduction in gel volume (McBride, 1978) or due to the poor symmetry of the binding sites (Bloom and McBride, 1979). The similarity of the H-peat and humic acid spectra to the pectin and IRC-50 spectra suggests that the Mn^{2+} in humic acid and peat Mn^{2+} may be bound by COO^- groups that are on nonadjacent carbons (Fig. 5).

Carboxylic acids form outer sphere complexes with monovalent ions, (Gordon, 1975) and it is likely that most divalent ions, with the exception of Cu^{2+}, form outer sphere complexes with acetic acid (Archer and Monk, 1964). The retention of water between COO^- and a metal ion is due to the hydrogen bonding of water to COO^- (Gordon, 1975) and the hydration energy of the cation. The predominance of outer sphere complexing need not be absolute because of the equilibrium:

$$
\begin{array}{ccccc}
\text{inner sphere} & \leftarrow & \text{outer sphere} & \leftarrow & \text{free} \\
\text{complex} & \rightarrow & \text{complex} & \rightarrow & \text{ion}
\end{array}
$$

The ESR spectra of Lakatos et al. (1977b) for Mn^{2+} in ovendried and wet humic acid were unlike those just discussed. The line widths measured were narrow, about 10 Gauss, much like that expected for Mn^{2+} doped into a solid state matrix. Lakatos et al. (1977b) concluded that Mn^{2+} forms an inner sphere complex with humic acid and the sites form a symmetrical environment for Mn^{2+}.

Several differences in experimental procedure could explain the differences in observations. Lakatos et al., (1977b) used low loading levels of Mn^{2+} in Na^+ humate. Thus, the pH at which the complex formed was high, possibly high enough to cause ionization of phenolic protons or formation of $MnOH^+$ ions. McBride (1978) and Bloom and McBride (1979) worked at pH values of 7 and below. Also, Lakatos et al. (1977b) removed more of the residual Fe^{3+} and Al^{3+} in their humic acid. These ions may block some sites which have a greater affinity for cation binding.

The ESR spectra for Cu^{2+} on H-peat (Fig. 4) are similar to Cu^{2+} in humic acid (McBride, 1978). Because of Jahn-Teller distortion (Huheey, 1972), the d electrons in Cu^{2+} are not spherically symmetrical and ligands surrounding Cu^{2+} are not in perfect octahedral symmetry. Rather, one axis is elongated relative to the other two. In Fig. 4b, two different components are observed, with different g-values, indicating that Cu^{2+} is immobilized on binding sites. One component (g ∥) appears when the magnetic field is parallel to the elongated (z) axis while the other (g ⊥) appears when the field is perpendicular to the z-axis. Similar g values and hyperfine splitting have been observed for Cu^{2+} in pectin and Amberlite IRC-50 (Lakatos et al., 1977b). The broadening of the spectrum and the small splitting constant for g ⊥ obscures much of the detail. The solution spectrum (Fig. 4a) results from an average of the g ⊥ and g ∥ components by the freely tumbling ions. A broad resonance at about g = 2.18, similar to the solution spectrum, appears in Fig. 4b but largely disappears upon drying (Fig. 4c). This phenomenon, also observed in humic acid (McBride, 1978), suggests that some Cu^{2+} ions are free to tumble but this mobility is lost when the ion is dehydrated.

The restriction of Cu^{2+} mobility, while indicating inner sphere complex formation, is not necessarily indicative of chelate formation (McBride, 1978). The spectra do indicate that N-containing sites are not involved with binding since the g values are similar to that of hydrated Cu^{2+} (McBride, 1978). It is likely that both water and carboxyl oxygens form coordinate bonds with Cu^{2+} (Fig. 6) and that the charge on some carboxyl ions is shared by two (or more) metal ions (Fig. 6). Lakatos et al. (1977b) concluded from Cu^{2+}-humic acid ESR spectra that Cu^{2+} forms an inner sphere complex that incorporates some N. These spectra were obtained using procedures discussed earlier and it is likely that the treatment differences contribute to the difference in Cu^{2+} spectra from those obtained by others (McBride, 1978; Bloom and McBride, 1979).

Additional ESR evidence indicates that N-containing ligand groups may bind small quantities of Cu^{2+}. Copper (II) porphyrin ESR spectra have been observed both in peat (Goodman and Cheshire, 1973) and in a mineral soil (Cheshire et al., 1977). The amount of Cu involved may be only a small fraction of the total humate Cu (Cheshire et al., 1977).

The vanadyl ion, VO^{2+}, which is a form of vanadium found in soils (Cheshire et al., 1977) has also been used in ESR metal ion binding studies. Like Cu^{2+} it is strongly bound and forms an inner sphere complex in peat (Goodman and Cheshire, 1975) and humic acids (McBride, 1978; Lakatos et al., 1977a). The spectra obtained for VO^{2+} in humic acid (McBride, 1978) and in peat (Goodman and Cheshire, 1975) are similar to that of VO^{2+} in pectin (Lakatos et al., 1977b). The spectra of VO^{2+} in humic acid obtained by Lakatos et al. (1977b) were somewhat different. The difference from the pectin spectrum was attributed to phenolic OH participation in the complex.

1977a). From this similarity, Hansen and Mosback (1970) concluded that Fe^{3+} in fulvic acid also is at the apices of an equilateral triangle bridged by COO^-. Such detailed structural inferences cannot be made from Mössbauer data alone but it is likely that Fe^{3+} carboxyl bridges occur and it may be that phenolic oxygens contribute to binding (Lakatos et al., 1977a).

Computer techniques have been used to separate the Mössbauer spectrum of Fe^{3+} on humic acid into three components (Senesi et al., 1977), each attributed to Fe binding at a different site. One site had parameters similar to that of Fe^{2+} which suggest the possibility of reduction of some Fe^{3+}. Some of the Mössbauer absorption may have been due to mineral iron since one of the Fe^{3+} components was not decreased in intensity by hydrazine reduction.

The Mössbauer spectra of Fe^{2+} in humic acid after drying at 130 C are similar to the spectrum of Fe^{2+} in a chelating resin (Lakatos et al., 1977a). This suggests that Fe^{2+} forms an inner sphere complex in the dried state. The spectra of solution Fe humate (frozen) indicates that the Fe^{2+} ions are at least partially hydrated. Lakatos et al., (1977a) concluded that Fe^{2+} formed an inner sphere complex but retained some hydration water. Freezing, however, lowers the chemical potential of water and is equivalent to drying. It may be that under truly wet conditions Fe^{2+} is completely hydrated.

RELATIVE PREFERENCE FOR METAL ION BINDING

Alkaline Earth and Alkali Metals

Preference for binding of various metal ions by organic matter has been studied using methods developed for the study of soluble metal-ion complexes and metal ion binding in polyelectrolytes (Schnitzer and Khan, 1972; Stevenson and Ardakani, 1972; and Jellinek, 1974). Many different methods have been used and it is often difficult to compare the results of one investigation with those of another.

Ion exchange studies with alkaline earth and alkali metals suggest that soil organic matter behaves like a weak field electrolyte. Studies with K-humate, Ca-humate and H-peat show that the preference for alkali metals is rubidium (Rb^+) > cesium (Cs^+) > potassium (K^+) > sodium (Na)$^+$ > lithium (Li^+) and for alkaline earths is barium (Ba^{2+}) > strontium (Sr^{2+}) > calcium (Ca^{2+}) > magnesium (Mg^{2+}) (Zadmard, 1939; Bel'kevich et al., 1973). This is similar to the preference shown by linear polysulfonates (Reichenberg, 1966) and permanent charge clay minerals (Van Olphen, 1977). Fulvic acids show a preference for Ca^{2+} over Mg^{2+} (Schnitzer and Hansen, 1970) but prefer Na^+ over K^+ (Gamble, 1973).

Weak field behavior is not expected for a polycarboxylic acid. Monomeric carboxylic acids are strong field and show the inverse relative preference for metal ions compared to weak field exchangers. Polycarboxylates, however, exhibit variable orders (Gordon, 1975). Three different carboxylate resins studied by Reichenberg (1965) show the following orders:

$$Li > Cs > Na > K$$
$$Li > Na > Cs > K$$
$$Li > Na > K > Cs$$

Thus, various orders are possible for synthetic polycarboxylic acids but none seem to exhibit a completely weak field behavior.

Weak field preference by humic acids is corroborated by coagulation data (Ong and Bisque, 1968). Also, like smectite clays, humic materials are flocculated at very low solution concentrations of polyvalent ions due to the bridging effects of polyvalent ions (Table 2).

Ion exchange data also show that M^{2+} ions are greatly preferred over M^+ ions. This, along with data that suggest the preferential binding of Al^{3+} by organic matter relative to smectite in a Al^{3+}-Ca^{2+} system, suggests that ion charge is a more important factor in determining binding in organic matter than it is in clays (Bloom and McBride, 1979). In soil, the M^{3+} saturation of organic matter sites is likely greater than that for clays. For M^+ ions the situation will likely be the inverse.

Dissimilar behavior of alkali metal cations bonded to organic matter, indicates that there is a close association between binding sites and the hydrated M^+ cations. Alkali metal ions in the vicinity of a charged polymer surface can be separated into two populations; a diffuse counter ion layer and ions "condensed" on the surface in the Stern layer. The fraction of sites neutralized by the condensed cations is a function of charge density and at higher charge density can be $> 90\%$ (Manning, 1977). Studies with synthetic polycarboxylates show that the fraction of sites neutralized varies from 0.8 to 0.9 for a cross-linked gel (Gustafson, 1963) and 0.6 to 0.8 for a linear polymer (Alexandrowicz, 1959). Ion binding experiments of Gamble (1973) suggest Na^+ and K^+ condense on fulvic acid.

Potentiometric Titration

Potentiometric titrations have been used to study cation binding in organic matter. Titration curves of humic fulvic acids in alkali metal and

Table 2. Critical concentrations for the coagluation of sodium humate at pH 7 (Ong and Bisque, 1968).

Critical concentrations for coagulation	
	mmoles/1
LiCl	826
NaCl	598
KCl	335
$MgCl_2$	30
$CaCl_2$	7.2
$AlCl_3$	0.50
$FeCl_3$	0.64

alkaline earth salts have a characteristic sigmoidal shape (Fig. 7). The inflection indicated by the point of maximum slope (arrow Fig. 7) is usually taken to be the end point for the titration of COOH. In humic acid this occurs beween pH 7 and 8 (Posner, 1964; Stevenson, 1976; Takamatsu and Yoshida, 1978). The quantity of acidity titrated depends on the humic acid, the cation of the salt solution and the salt concentration. Increasing the salt concentration shifts the curve to lower pH values. The end point for the titration of phenolic protons is obscured by the high concentration of free OH^- necessary to attain the end point pH.

Studies of titration behavior suggest that soil organic matter may contain three types of weak acid protons, a very weak acid phenolic proton and two carboxyl protons with similar acidities. The titrations of organic matter, however, do not show the clear separation of carboxyl protons that is seen in the titration of the polymer formed from ethylene maleic

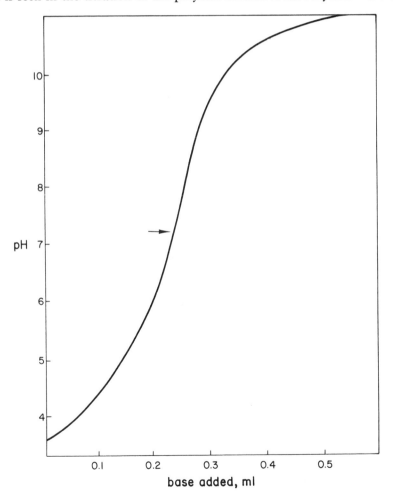

Fig. 7. Humic acid titrated with alkali metal hydroxide.

anhydride, a polymer which contains dicarboxyl sites with the carboxyl groups on adjacent carbons. Gamble (1972) suggested that in fulvic acid one of the carboxyl groups is ortho to a phenolic –OH group. Choppin and Kullberg (1978) suggested that humic acid contains benzoate, salicylate, and phenolate groups while Perdue (1978) suggested that at least one-third of the COOH groups are not ortho to OH groups.

Titration data for soil organic matter fit the generalized Henderson-Hasselbalch equation:

$$pH = pK - n \log \left(\frac{1-\alpha}{\alpha} \right) \qquad [1]$$

where α is the extent of proton dissociation and n and pK are empirical constants. For a monocarboxylic acid $n = 1$ and pK is the negative log of the acid dissociation constant. In humic acid and H-peat n approximates 2 (Posner, 1964; Bloom and McBride, 1979). With increasing salt concentration n is invariant but pK decreases. Similar behavior has been shown for cross-linked polycarboxylic acids (Gregor et al., 1955a; Fisher and Kunin, 1956). In linear polycarboxylic acids, n decreases with increasing ionic strength (Gregor et al., 1955b).

As the degree of cross-linking increases, CEC decreases because cross-links carry no functional groups. An increase in cross-linking also results in pK increase (Fig. 8) (Fisher and Kunin, 1956). Humic acid fractions extracted by different methods from the same soil show (Fig. 9) a similar negative correlation between pK and CEC. The difference in the acidities of humic materials may relate charge density and cross-linking rather than pK_a values of different sites as proposed by Posner (1964, 1966).

Titration curves for organic matter in salt solutions of various metal ions demonstrate directly the relative ability of metal ions to replace protons (Fig. 10 and Fig. 11). Cations that hydrolyze readily cause an apparent shift in the end points (Fig. 10) and a second buffer region (Fig. 11). This is due to the hydrolysis of these ions on the exchange surface and to the precipitation of the metal hydroxides (Van Dijk, 1971; Schnitzer and Khan, 1972; Bloom et al., 1979; Bloom and McBride, 1979).

Relative binding strengths for divalent ions can be compared to the binding strength of Ca^{2+} by calculating the ratio of Henderson-Hasselbalch constants (Fig. 12). H-peat shows little preference, compared to Ca^{2+}, for the binding of Mn^{2+} and Ni^{2+} nor does humic acid bind Mn^{2+}, Fe^{2+}, Co^{2+}, Ni^{2+}, and Zn^{2+} preferentially to Ca^{2+}. This is unlike the preference series obtained for fulvic acid by Schnitzer and Hansen (1970). The greater charge density of fulvic acid may account for some of the difference in behavior. The similarity of the ions listed above to Mn^{2+} indicates that these ions, like Mn^{2+}, are bound to peat and humic acid largely in outer sphere complexes. The greater strength of the Cu^{2+} binding is expected from the ESR data which indicates Cu^{2+} forms inner sphere complexes with peat and humic acid. The similarity in the binding strength of Pb^{2+} to that of Cu^{2+} (Stevenson, 1976) indicates that it also forms an inner

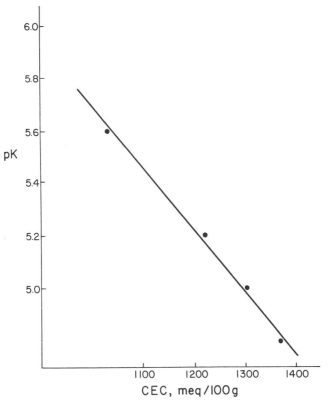

Fig. 8. The relationship between CEC and pK values, determined from Eq. 1, for cross-linked polycarboxylic acids with different fractions of cross-linking.

sphere complex. From the above data the following preference series for divalent ions can be proposed for humic acids and peat:

$$Cu > Pb \gg Fe > Ni = Co = Zn > Mn = Ca$$

From Fe to the right little difference in preference is shown. More data is needed to establish the exact relative preference of these ions.

CONCLUSIONS

Much of the evidence concerning the nature of binding sites and cation association with these sites is contradictory. Much more work will have to be completed before a detailed model of binding can be described with a high degree of confidence. Several general conclusions, however, can be drawn from the evidence. At pH values in most soils (<8) the cation binding behavior of organic matter can be modelled well by the be-

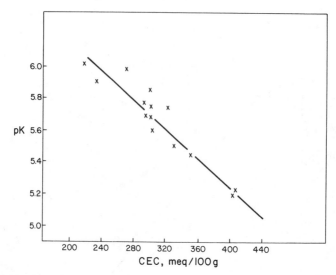

Fig. 9. The relationship between CEC and pK values, determined from Eq. 1, for humic acids extracted from the same soil using different extraction methods (Posner, 1966).

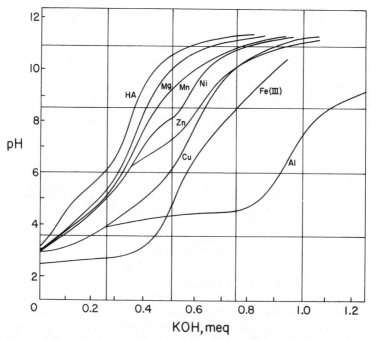

Fig. 10. Titration of 0.50 meq of humic acid in a solution containing 0.50 meq of metal ion (Van Dijk, 1971).

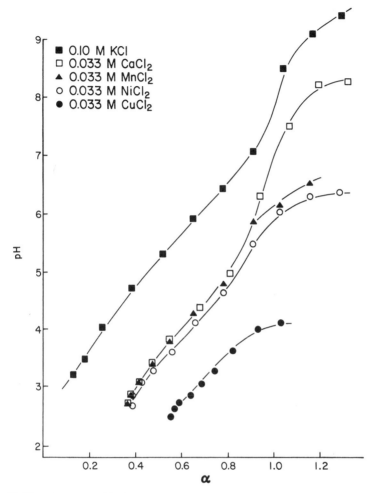

Fig. 11. Titration curves of 0.32 meq H^+-peat in 50 ml of chloride salts of K^+, Ca^{2+}, Mn^{2+}, Ni^{2+}, and Cu^{2+} showing the relationship between fractions of sites occupied by metal ions, α, and pH (Bloom and McBride, 1979).

havior of natural and synthetic polycarboxylic acids with COOH groups on nonadjacent carbons. In aqueous suspension, a large fraction of monovalent ions is condensed on the surface (i.e. in the Stern layer), neutralizing charge. Most divalent ions, with the exception of Cu^{2+}, Pb^{2+}, and VO^{2+}, are bound in outer sphere complexes. The more tightly bound ions while largely inner sphere, are likely partially hydrated. Chelation need not be postulated to describe binding; however, chelation sites may exist as minor components. These sites may be blocked by Fe^{3+} and/or Al^{3+} in soil.

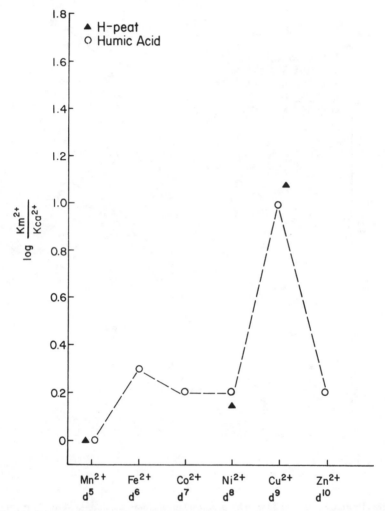

Fig. 12. Equation 1 binding constants for d^5 to d^{10} divalent ions relative to the constants for Ca^{2+} on H^+-peat and humic acid. H-peat constants calculated from Fig. 11 and humic acid constants estimated from pH drop for metal ion addition at $\alpha = 1/2$ (Bloom and McBride, 1979).

Nitrogen containing sites (e.g. porphyrins) make a minor contribution to the binding of Cu^{2+} and possibly other metal ions. There is no evidence of binding by sulfur containing sites.

LITERATURE CITED

1. Alexandrowicz, A. 1959. Activities of polyelectrolytes. J. Polymer Sci. 40:91–106.
2. Anderson, G. 1975. Sulfur in soil organic substances. *In* J. E. Gieseking (ed.) Soil components: Vol. 1, Organic components. Springer Verlag, New York.

3. Archer, D. W., and C. B. Monk. Ion association constants of some acetates by pH (glass electrode) measurements. J. Chem. Sco. 1964:3,117–3,122

4. Banerjee, S. K., and S. K. Mukherjee. 1972. Studies on the infrared spectra of some divalent transitional metal humates. J. Ind. Soc. Soil Sci. 20:91–94.

5. Bel'kevich, P. I., L. M. Rogach, and L. R. Christova. 1973. Peat selectivity in relation to alkali metal and alkaline earth ions. Vestsi Akad Nauuk Belarus SSR, Ser. Khim. Nauuk 1973:35–38.

6. Bloom, P. R., M. B. McBride, and R. M. Weaver. 1979a. Aluminum organic matter in acid soils: Buffering and solution aluminum activity. Soil Sci. Soc. Am. J. 43:488–493.

7. ————, ————, and ————. 1979b. Aluminum organic matter in acid soils: Salt-extractable aluminum. Soil Sci. Soc. Am. J. 43:813–815.

8. ————, and ————. 1979. Metal ion binding and exchange with hydrogen ions in acid-washed peat. Soil Sci. Soc. Am. J. 43:687–692.

9. Broadbent, F. E., and G. R. Bradford. 1952. Cation exchange groupings in the soil organic fraction. Soil Sci. 74:447–457.

10. Cheshire, M. V., M. L. Berrow, B. A. Goodman, and C. M. Mundie. 1977. Metal distribution and nature of some Cu, Mn, and V complexes in humic and fulvic fractions of soil organic matter. Geochimica Cosmochimica Acta. 41:1,131–1,138.

11. Choppin, G. R., and L. Kullberg. 1978. Protonation thermodynamics of humic acid. J. Inorg. Nucl. Chem. 40:651–654.

12. Davies, R. I., M. V. Cheshire, and I. J. Graham-Bryce. 1969. Retention of low levels of copper by humic acid. J. Soil Sci. 20:65–71.

13. Fisher, S., and R. Kunin. 1956. Effect of cross-linking on the properties of carboxylic polymers: I. Apparent dissociation constants of acrylic and methacrylic acid polymers. J. Phys. Chem. 60:1030–1032.

14. Flaig, W., H. Beutelspacher, and E. Rietz. 1975. Chemical composition and physical properties of humic substances. In J. E. Gieseking (ed.) Soil components: Vol. 1, organic components, Springer Verlag, New York.

15. Gamble, D. S. 1972. Potentiometric titration of fulvic acid: Equivalence point calculations and acidic functional groups. Can. J. Chem. 50:2680–2690.

16. ————. 1973. Na$^+$ and K$^+$ binding by fulvic acid. Can. J. Chem. 51:3217–3222.

17. ————, C. H. Langford, and J. P. K. Tong. 1976. The structure and equilibrium of a manganese II. complex of fulvic acid studies by ion exchange and nuclear magnetic resonance. Can. J. Chem. 54:1,239–1,245.

18. Geering, H. R., J. F. Hodgson, and C. Sdano. 1969. Micronutrient complexes in soil solution: IV. The chemical state of manganese in solution. Soil Sci. Soc. Am. Proc. 33: 81–85.

19. Gillam, W. S. 1940. A study on the chemical nature of humic acid. Soil Sci. 49:433–453.

20. Giovannini, G., and P. Sequi. 1976. Iron and aluminum as cementing substances of soil aggregates. II. Changes in stability of soil aggregates following extraction of iron and aluminum by acetyl acetone in a non-polar solvent. J. Soil Sci. 27:148–153.

21. Goodman, B. A., and M. V. Cheshire. 1973. Electron paramagnetic resonance evidence that copper is complexed in humic acid by prophyrin groups. Nature New Biol. 244: 158–159.

22. ————, and ————. 1975. The bonding of vanadium in complexes with humic acid: An electron paramagnetic resonance study. Geochim. Cosmochim. Acta. 39:1711–1713.

23. Gordon, J. E. 1975. The organic chemistry of electrolyte solutions. John Wiley and Sons, New York.

24. Gregor, H. P., M. J. Hamilton, J. Becher, and F. Bernstein. 1955a. Studies on ion exchange resins. XIV. Titration capacity and swelling of methacrylic resins. J. Phys. Chem. 59:874–881.

25. ————, L. B. Luttinger, and E. M. Loebl. 1955b. Meta-polyelectrolyte complexes: II complexes of Cu(II) with cross-linked polycarboxylic and polymethacrylic acid. J. Phys. Chem. 59:366–368.

26. Gustafson, R. L. 1963. Donnan equilibrium on cross-linked polymethacrylic acid-sodium chloride systems. J. Phys. Chem. 67:2,549–2,557.

27. Hansen, E. H., and H. Mosback. 1970. Mössbauer studies of an iron III. fulvic acid complex. Acta. Chem. Scand. 24:3,083–3,084.

28. Hayes, M. H. B., and R. S. Swift. 1978. The chemistry of soil organic colloids. p. 179–320. In D. J. Greenland and M. H. B. Hayes, The chemistry of soil constituents. John Wiley and Sons, New York.

29. Helling, C. S., G. Chesters, and R. B. Corey. 1964. Contribution of organic matter and clay to soil cation-exchange capacity as affected by pH and the saturating solution. Soil Sci. Soc. Am. Proc. 28:517–520.

30. Huheey, J. E. 1972. Inorganic chemistry: Principles of structure and reactivity. Harper and Row, New York.

31. Jellinek, H. H. G. 1974. Soil organics I. Complexation of heavy metals. Cold Regions Res. Eng. Lab., Hanover, New Hampshire.

32. Lakatos, B., L. Korecz, and J. Meisel. 1977a. Comparative study on the Mössbauer parameters of iron humates and polyuronates. Geoderma. 19:149–157.

33. ⎯⎯⎯, T. Tibai, and J. Meisel. 1977b. EPR spectra of humic acids and their metal complexes. Geoderma 19:319–338.

34. Manning, G. S. 1977. Limiting laws and counter ion condensation in polyelectrolyte solutions. IV. The approach to the limit and the extraordinary stability of the charge fraction. Biophysical Chem. 7:95–102.

35. Martin, A. E., and R. Reeve. 1958. Chemical studies of podzolic illuvial horizons III. Titration curves of organic matter suspensions. J. Soil Sci. 9:89–100.

36. McBride, M. B. 1978. Transition metal bonding in humic acid: An ESR study. Soil Sci. 126:200–209.

37. ⎯⎯⎯, T. T. Pinnavaiia, and M. M. Mortland. 1975. Electron spin relaxation and the mobility of manganese II. Exchange ion in smectites. Am. Mineral 60:66–72.

38. McCarthy, P., and H. B. Mark, Jr. 1975. Infrared studies on humic acid in deuterium oxide: I. Evaluation and potentialities of the technique. Soil Sci. Soc. Am. Proc. 39:663–668.

39. ⎯⎯⎯, ⎯⎯⎯, and P. R. Griffiths. 1975. Direct measurement of infrared spectra of humic substances in water by Fourier transform infrared spectroscopy. J. Agric. Food Chem. 23:600–601.

40. Ong, H. L., and R. E. Bisque. 1968. Coagulation of humic colloids by metal ions. Soil Sci. 106:220–224.

41. Perdue, E. M. 1978. Solution thermochemistry of humic substances. I. Acid-base equilibria of humic acid. Geochim. Cosmochim. Acta. 42:1351–1358.

42. Posner, A. M. 1964. Titration curves of humic acids. 8th Int. Congr. Soil Sci. 17:161–173.

43. ⎯⎯⎯. 1966. The humic acid extracted by various reagents from soil. Part I. Yield, inorganic components, and titration curves. J. Soil Sci. 17:65–78.

44. Reichenberg, D. 1966. Ion exchange selectivity. p. 227–278. In J. A. Marinsky (ed.) Ion exchange, Vol. 1. Marcel Dekker, New York.

45. Senesi, N., S. M. Griffith, and M. Schnitzer. 1978. Binding of Fe^{3+} by humic materials. Geochim. Chemochim. Acta. 41:969–976.

46. Schnitzer, M. 1978a. Some observations on the chemistry of humic substances. Agrochim. 22:216–225.

47. ⎯⎯⎯. 1978b. Humic substances: Chemistry and reactions. p. 1–58. In M. Schnitzer and S. U. Khan (ed.) Soil organic matter. Elsevier, New York.

48. ⎯⎯⎯, and S. I. M. Skinner. 1965. Organo-metallic interactions in soils: 4. Carboxyl and hydroxyl groups in organic matter and metal ion retention. Soil Sci. 99:278–284.

49. ⎯⎯⎯, and S. U. Khan. 1972. Humic substances in the environment. Marcel Dekker, New York.

50. ⎯⎯⎯, and E. M. Hansen. 1970. Organo-metallic interactions in soils: 8. An evaluation of methods for the determination of stability constants of metal fulvic acid complexes. Soil Sci. 109:333–340.

51. Stevenson, F. J. 1976. Stability constants of Cu^{2+}, Pb^{2+}, and Ca^{2+} complexes with humic acids. Soil Sci. Soc. Am. J. 40:665–672.

52. ————, and M. S. Ardakani. 1972. Organic matter reactions involving micronutrients in soils. p. 79–114. In J. J. Mortvedt, P. M. Giordano, and W. L. Lindsay (ed.) Micronutrients in agriculture. Soil Sci. Soc. Am., Madison, Wis.

53. ————, and K. M. Goh. 1971. Infrared spectra of humic acids and related substances. Geochim. Cosmochim. Acta. 35:471–483.

54. ———— and ————. 1974. Infrared spectra of humic acids: Elimination of interferences due to hygroscopic moisture and structural changes accompanying heating with KBr. Soil Sci. 117:34–41.

55. Takamatsu, T., and T. Yoshida. 1978. Determination of stability constants of metal humic acid complexes by potentiometric titration and ion selective electrodes. Soil Sci. 125:377–386.

56. Tan, K. H. 1977. Infrared spectra of fulvic acids containing silica metal ions and hygroscopic moisture. Soil Sci. 123:235–240.

57. Van Dijk, J. 1971. Cation binding of humic acids. Geoderma 5:53–67.

58. Van Olphen. 1977. an introduction to clay colloid chemistry, 2nd ed. John Wiley and Sons, New York.

59. Vinkler, P., B. Lakatos, and J. Meisel. 1976. Infrared spectroscopic investigations of humic substances and their metal complexes. Geoderma 15:231–242.

60. Wagner, G. H., and F. J. Stevenson. 1965. Structural arrangement of functional groups in soil humic acid as revealed by infrared analyses. Soil Sci. Soc. Am. Proc. 29:43–48.

61. Wilson, S. A., and J. H. Weber. 1977. Electron spin resonance analysis of semiquinone free radicals of aquatic and soil fulvic and humic acid. Anal. Lett. 10:75–84.

62. Wilson, M. A., A. J. Jones, and P. Williamson. Nuclear magnetic resonance spectroscopy of humic materials. Nature 276:487–489.

63. Wood, J. C., S. E. Moschopedis, and W. Den Hertog. 1961. Studies in humic acid chemistry II. Humic anhydrides. Fuel 40:491–502.

64. Zadmard, H. 1939. Zur kenntnis den Kolloidchemiseen Eigenschaften des Humus. Kolloid Beihefte. 49:315–364.

QUESTIONS AND ANSWERS

Q. What are the relative positions of carboxylate and water oxygens in the coordination of copper by humates?

A. This question relates to Fig. 6. The picture presented is a very general one and is at present impossible to label the axes. The z-axis is elongated by the Jahn-Teller distortion. At first glance it seems possible to make a unique assignment of water or carboxylate groups to z-axis positions but this turns out to be a difficult problem. I've puzzled over this for a long time, but haven't come to any firm conclusions.

Q. Why do you postulate six-fold coordination for copper? Isn't it true that organic complexes of copper are usually square planar?

A. This is true for nitrogen containing ligands but not oxygen ligands. Typically, as in the hexa-aquo complex, oxygen ligands form tetragonal complexes. The ESR signal of copper on peat or humic acid is very much like that of the hexa-aquo complex of copper on hectorite.

Q. In your analysis of functional groups that would complex metals, you haven't mentioned the possible role of phenolic oxygen. Although phenolic oxygen groups may not be dissociated at high pH, they are easily oxidized at much lower pH's. For instance, hydroquinone becomes easily oxidized to quinone, gallic acid is easily oxidized, and ascorbic acid is easily oxidized. The resulting oxygen is something like carbonyl oxygen which could easily complex metals.

A. Clearly much of the color of humates is due to quinone and semiquinone type groups. The exact role of the OH groups in binding is unclear. All we say is that polycarboxylic acids seem to be very similar to humic substances with respect to cation binding. This doesn't necessarily mean, that the OH groups don't enter in, but recent work by M. B. McBride indicates that salicylic acid is not a good model for humic binding sites. More research is needed to better define the possible role of the OH groups.

Q. I think if you look at polyphenolic compounds you see different behavior. For instance, in gallic acids the three OH groups are much more acidic than salicylic acid or phenol itself.

A. Dr. J. P. Martin has been kind enough to offer to give me some of his polyphenolic polymers. I'm going to start working on those to see what they look like. This kind of research is new and we might have to revise some of our conclusions in the future.

Q. I was just wondering where cadmium fits in on this last slide you showed?

A. It is hard to rank the relative preference of ions without using the same method and humic materials for the whole series of ions. I intend in the future to do this with a large number of divalent ions including cadmium.

CHAPTER 8

Effect of Organic Matter on Exchangeable Aluminum and Plant Growth in Acid Soils[1]

W. L. HARGROVE AND G. W. THOMAS[2]

INTRODUCTION

Although J. T. Way did not recognize that soil organic matter exhibited cation exchange behavior in his original experiments (Way, 1850), it was found soon afterwards that organic matter had an even larger so-called "base-exchange capacity" than the mineral component of the soil (Johnson, 1859). In this century, McGeorge (1931) was one of the first soil scientists to quantitatively study soil organic matter in regard to cation exchange. He observed that, in general, divalent cations were held more tightly by humic acid than monovalent cations and, thereby, was first to

[1] Contribution from the Univ. of Kentucky, Lexington, KY 40546. Published with the approval of the Director of the Kentucky Agric. Exp. Stn. as J. article no. 79-3-150.

[2] Graduate research assistant and professor, respectively.

propose a lyotropic series for organic matter. He also made another important observation: leaching humic acid with strong HCl as a pretreatment increased the effective CEC. McGeorge did not know, however, exactly which functional groups were responsible for cation exchange and simply related CEC to the lignin content. Later, Gillam (1940) and Broadbent and Bradford (1952) showed that the main functional groups involved in cation exchange and in acidity in organic matter were carboxyl groups with some minor contributions from phenolic, enolic, and alcoholic hydroxyls. (For recent reviews on the chemical composition and structure of organic matter, see Allison, 1973; Flaig et al., 1975; and Schnitzer and Khan, 1972.)

Assuming the main source of acidity in organic matter to be carboxylic groups, titration curves of soil organic matter showed an abnormal weakness compared to titration curves of pure carboxylic acids. Most carboxylic acids have Ka's of 10^{-4} to 10^{-5} whereas titration curves of organic matter gave values of about 10^{-6}. Martin and Reeve (1958) showed that this abnormal weakness was due to Al and Fe present on exchange sites, just as the abnormal weakness of soil clays was caused by exchangeable Al^{3+} (Harward and Coleman, 1954) and even more so by hydroxylated Al (Coleman and Thomas, 1964). Removal of Fe and Al from organic matter gave an apparent Ka of 10^{-4}. Furthermore, Schnitzer and Skinner (1963a, 1963b) showed that Al was hydroxylated in organic matter with an average composition of $Al(OH)_2^+$ rather than being present as the neutral salt-exchangeable Al^{3+} ion as Chernov (1947) postulated in his treatise on soil acidity.

Several years before the work of Broadbent and Bradford (1952), Martin and Reeve (1958), or Schnitzer and Skinner (1964), Sante Mattson (1933) was studying the amphoteric nature of soils in relation to Al toxicity, and apparently recognized that plants growing in soils which were high in organic matter did not exhibit symptoms of aluminum toxicity at the same pH that soils low in organic matter did. Mattson demonstrated the preventive effect of organic matter against Al injury to plants by adding humic acid to an acid Nipe soil in a greenhouse experiment which is described in number XII of his series of some 30 papers on the laws of soil colloidal behavior (Mattson and Hester, 1933). Without humic acid added, he observed Al toxicity at pH values up to 4.8. However, with additions of humic acid, plants grew satisfactorily at pH 4.4. Later, Hester (1935) demonstrated the same effects of organic matter additions in relation to Al toxicity on three coastal plain soils in Virginia with three different vegetable crops.

At about this same time, X-ray diffraction became an analytical tool for soil chemists (Hendricks and Fry, 1930; Kelley et al., 1931), and apparently little attention was paid to Mattson's observations on the role of soil organic matter in Al toxicity. In the following 40 or more years most soil chemists concerned themselves with the mineral component of the soil, and as a result, knowledge of the nature of soil organic matter and of cation exchange and acidity in soil organic matter was overshadowed by new information and emphasis on crystalline clay mineral structures and their exchange properties.

Recently, we have been studying the effects of organic matter on Al equilibria in soils, prompted by the observations that surface soils often have lower exchangeable Al contents at pH values less than 5 than do subsoils (Coleman and Thomas, 1967) and that under no-tillage management, corn is able to yield satisfactorily even though the pH of the surface soil may get quite low as a result of high N fertilizer applications (Blevins et al., 1977).

In general, our objectives have been 1) to characterize Al-organic matter reactions in the laboratory using a well-humidified peat as a model for soil organic matter; and 2) to determine the effects of additions of peat to acid soils on Al equilibria and plant growth in the greenhouse. Peat seemed a good model for soil organic matter since it had a high CEC, and titrations of H-peat were similar to those reported for extracted soil organic matter. Stevenson (1976) also found that differences between humic acids extracted from surface soils, peat, and lignite coal in their ability to complex Cu^{2+}, Pb^{2+}, or Cd^{2+} were only slight.

CHARACTERISTICS OF ALUMINUM-ORGANIC MATTER REACTIONS

In earlier work on acidity of soil clays, titration curves of Al-clays showed a much weaker acid nature than H-clays (c.f. Harward and Coleman, 1954; Coleman and Thomas, 1964). This same property can be shown for Al-organic matter and H-organic matter. Figure 1 shows potentiometric titration curves for peat with various Al contents. The curves were obtained by batch titration of 1 g peat samples in 50 ml of $0.1N$ KCl with $0.1N$ KOH. The apparent pKa values for each of these peats shown in Fig. 1 were calculated using the Henderson-Hasselbalch equation:

$$pH = pKa + n \log \frac{\alpha}{100 - \alpha} \qquad [1]$$

where

n = constant (which = 1 for monoprotic acids such as acetic acid but is usually about 2 for polyprotic acids such as soil organic matter) (Martin and Reeve, 1958).

α = % dissociation of the organic acid.

Assuming 100% dissociation at pH 7, the pKa can be obtained from the pH at which the functional groups in the peat are 50% dissociated, since at 50% dissociation $n \log [\alpha/(100 - \alpha)] = 0$ and pH = pKa. For example, for titration curve f in Fig. 1, 0.40 meq OH^-/g peat was required to raise the pH to 7. The pKa therefore is equal to the pH after 0.20 meq OH^-/g was added, or pKa = 6.2.

As seen in Fig. 1, pKa values increased with increasing Al contents in the peat, demonstrating the weaker acid nature of the peat as the Al content increases. This same relationship was shown by Martin and Reeve (1958) for humic acid extracted from spodic horizons.

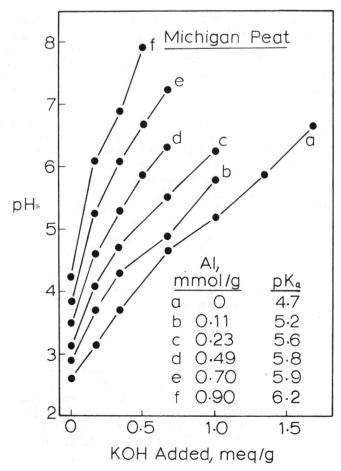

Fig. 1. Potentiometric titration curves and apparent pKa values for Michigan peat of varying
Al contents.

The amount of Al adsorbed on the peat is of course dependent on pH
since the dissociation of functional groups in organic matter is pH dependent. Figure 2 shows the amount of Al assumed to be adsorbed by peat as
a function of the suspension pH. This curve was obtained by equilibrating
1.0 mmol of AlCl₃ with 1 g of H-peat and various amounts of KOH in 50
ml of 0.1N KCl. After equilibration, the Al remaining in solution was determined by atomic absorption spectroscopy, and the Al not recovered
was assumed to be adsorbed by the peat. The amount of Al assumed to be
adsorbed by peat increased linearly with increases in pH over the range of
2.0 to 4.5, with almost all of the added Al being adsorbed at pH 4.5 (Fig.
2). It is doubtful that precipitation of Al from solution occurred over this
pH range. Mixtures of AlCl₃ and KOH equilibrated without peat at pH
values of 4.0, 4.2, and 4.5 resulted in 100% recovery of Al in solution and
no visual indications of precipitation (cloudiness) after 24 hours, even

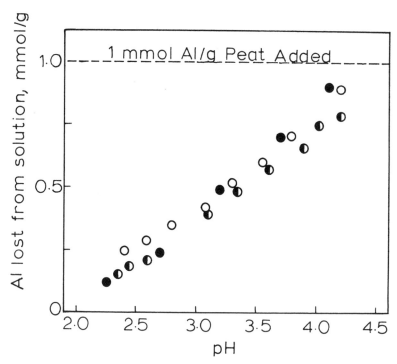

Fig. 2. Amount of Al assumed to be adsorbed by 1 g of H-peat in 50 ml of 0.1N KCl as a function of the solution pH. Results are from triplicate trials denoted by the three different symbols.

though cloudiness was apparent at the time of adding base. A pH of 4.5 corresponded to a OH/Al ratio of 2.25. At a OH/Al ratio of 2.50 and a pH value of 4.65, cloudiness persisted after 24 hours indicating precipitation. However, as seen in Fig. 2 the highest pH attained in the Al-peat systems was less than 4.5.

Not only does the amount of Al adsorbed by the peat increase with pH, but also the degree of hydrolysis, or the OH/Al ratio, of the adsorbed Al increases with pH. Figure 3 shows the relationship between the estimated OH/Al ratio and the pH of the solution in which the Al is adsorbed. The OH/Al ratio was estimated by using the titration curve of H-peat, the pH of the Al-peat in 0.1N KCl (after being washed free of excess $AlCl_3$ and resuspended in 0.1N KCl; note that this is not the same as the pH at which Al was adsorbed which is plotted in Fig. 3) and the Al content of the peat. The pH of the Al-peat was used to determine the amount of H^+ displaced from the H-peat titration curve. Dividing the meq H^+ displaced by the mmol Al adsorbed resulted in the average effective valence of the Al. From the average effective valence of the Al, the OH/Al ratio was determined. For example, referring to Fig. 1, a rise in pH of the Al-peat from 2.65 (for H-peat) to 3.15 (curve C) represented a loss of 0.20 meq H^+/g peat (obtained by determining the amount of base required to titrate

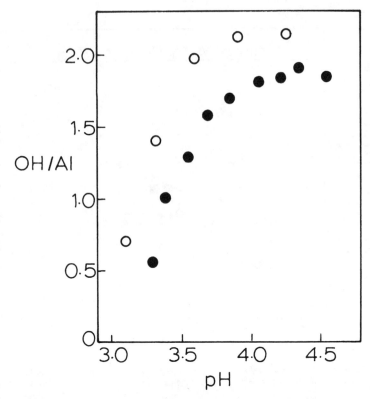

Fig. 3. Estimated OH/Al ratio of adsorbed Al as a function of solution pH. Results are from
   duplicate determinations denoted by different symbols.

the H-peat to the pH of the Al-peat). If the Al content was 0.23 mmol/g,
then the effective valence of the Al can be calculated:

$$\frac{0.20 \text{ meq H}^+ \text{ displaced}}{0.23 \text{ mmol Al adsorbed}} = 0.87^+ \qquad [2]$$

Thus, the average OH/Al ratio is 2.13.

   In Fig. 3 the estimated OH/Al ratio quickly approached 2.0 as the
pH at which Al adsorption occurred increased from 3.0 to 4.0. In contrast, in pure clay mineral systems Al is generally present as the $Al(H_2O)_6^{3+}$
ion over this pH range and is not appreciably hydrolyzed (Coleman and
Thomas, 1967). Apparently, Al is hydrolyzed to a greater extent over this
pH range in the presence of organic matter because the hydrolytic
product reacts with peat more extensively than with clay and because the
organic matter serves as a sink for hydrogen ions produced by hydrolysis
according to the following reactions:

$$Al(OOCR)_3 + H_2O \rightarrow Al(OH)(OOCR)_2 + RCOOH \qquad [3]$$

$$Al(OH)(OOCR)_2 + H_2O \rightarrow Al(OH)_2(OOCR) + RCOOH \qquad [4]$$

This helps explain Fig. 2 also, since with an increase in the OH/Al ratio each mole of adsorbed Al counters less negative charge in organic matter, freeing additional functional groups for more Al adsorption. At the same time the number of ionized sites is also increasing as the pH increases from 3.0 to 4.0 allowing more Al adsorption. Schnitzer and Skinner (1964) also calculated the average OH/Al ratio for extracted soil organic matter with varying Al contents and found that for low amounts of complexed Al ($\leq 5\%$ by weight) the OH/Al ratio was 1, but for moderate to high amounts (10–15% by weight) it approached 2.

In addition, the Al held by organic matter seems to be quite stable as an Al-organic matter complex since the Al is not exchangeable with KCl, but apparently is titratable (McLean et al., 1965). However, in our experiments a large portion of the Al could be removed with strong (2N) HCl, and recently Juo and Kamprath (1979) have used $CuCl_2$ solutions to partially extract Al from organic matter. Since the bound Al is not very exchangeable, however, the effective CEC of organic matter is inversely proportional to the amount of Al present. This relationship can be seen best in the data of Posner (1966). Since it has been shown in many cases that humic and fulvic acids extracted from acid soils do not appear to have many free –COOH groups but are present largely as Fe and Al complexes (Martin and Reeve, 1958; Bhumbla and McLean, 1965; Schnitzer and Gupta, 1964; and Griffith and Schnitzer, 1975), organic matter may contribute very little to the "effective" CEC of many soils.

RELATIONSHIP BETWEEN ORGANIC MATTER CONTENT AND EXCHANGEABLE ALUMINUM

The effect of organic matter on KCl exchangeable Al is evident in the data shown by Coleman and Thomas (1967). The KCl-exchangeable Al as a fraction of the CEC was less for surface soils from Virginia than for subsoils. Later, Thomas (1975) showed that KCl-exchangeable Al was lower at any given pH as organic matter increased in Maury silt loam. Evans and Kamprath (1970) and Clark and Nichol (1966) have also found much less exchangeable Al in organic soils than in mineral soils, even though the pH of the organic soils was quite low.

To determine the effect of organic matter additions on aluminum equilibria, 10 g of Zanesville subsoil (Typic Fragiudalf) was equilibrated in 25 ml of 0.1N KCl with various amounts of H-peat and 0.1N KOH. After equilibrating for 24 hours, the pH and Al concentration were determined in the filtered solutions. Some results of these experiments are shown in Fig. 4. The effect of adding H-peat was to decrease the amount of Al in solution at any given pH with increasing peat additions. Even as little as 2.5% peat by weight made a substantial difference in the amount of Al in solution at a given pH.

Hoyt and Turner (1975) also observed a decrease in soluble Al in acid soils to which fresh alfalfa (*Medicago sativa* L.) meal had been added. However the effect was only temporary, and the amount of exchangeable Al returned to the original amount after about 6 months.

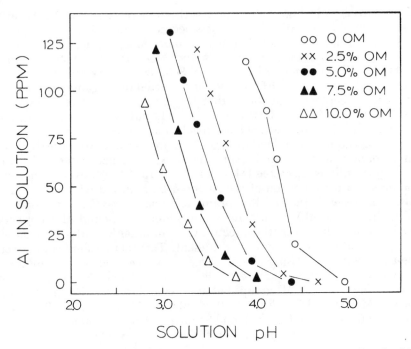

Fig. 4. The concentration of Al in solution after equilibrating 10 g of Zanesville subsoil in
0.1N KCl with various amounts of H-peat and base as a function of solution pH.

As seen in Fig. 4, small increases in organic matter can cause substantial decreases in the amount of Al in solution at a given pH. This
implies that organic matter may be important in preventing Al toxicity to
plants by lowering the amount of Al in solution, and that plants might
grow satisfactorily at a lower pH with additions of organic matter.

RELATIONSHIP BETWEEN ORGANIC MATTER CONTENT
AND PLANT GROWTH WITH RESPECT TO
ALUMINUM TOXICITY

To test the idea that organic matter can reduce Al toxicity we conducted a greenhouse experiment in which H-peat was added to a mixture
of sand and Al-montmorillonite. The montmorillonite was commercial
grade "vol-clay" and was saturated with Al by leaching with $1M$ $AlCl_3$
and then washing until free of Cl^-. The H-peat was prepared by leaching
commercial grade Michigan peat with $2N$ HCl and washing until free of
Cl^-. When Al^{3+}-saturated, the clay had 68 meq exchangeable Al^{3+}/100 g
clay, and when H^+-saturated, the peat had a CEC of 175 meq/100 g at pH
7. The sand-clay mixture was 5% clay by weight, and half of the pots received enough H-peat to make the mixture 10% organic matter by
weight. Lime was applied at nine different rates and was mixed uniformly with the soil. Barley (Hordeum vulgare var. Kearney) was planted in

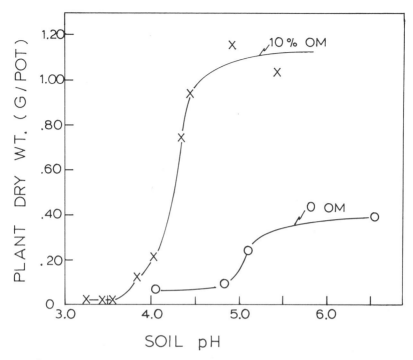

Fig. 5. Plant top dry weight as a function of soil pH for a sand + Al-montmorillonite mixture (0 OM) and a sand + Al-montmorillonite + H-peat mixture (10% OM).

each pot, watered daily with 1/2-strength Hoagland solution, and harvested after 24 days. The Kearney variety is known to be Al-sensitive (Reid, 1976). Each treatment was replicated three times.

The dry matter produced in those 24 days is shown in Fig. 5 as a function of the soil pH. The best yield of dry matter occurred at a lower pH with the addition of peat than with no peat added. Dry matter production was good at about pH 4.2 and above in the pots which contained 10% peat as opposed to about pH 5.0 and above for pots with no peat added. In fact, at any given pH, growth was much better with organic matter added. Similar results have been obtained by Bartlett and Riego (1972). They grew plants by solution culture in the presence of Al-citrate, Al-EDTA, Al-fulvate, Al(OH)$_2$Cl, and no Al. There was no difference in plant growth between plants growing in the presence of Al-citrate, Al-EDTA, Al-fulvate, or no Al. However, plant growth in solutions containing Al(OH)$_2$Cl was much worse.

In our study, yields for the sand-clay mixture without H-peat (Fig. 5) were relatively low even when the pH was sufficiently high to prevent Al toxicity. Even though all pots were watered with 1/2 Hoagland solution, there was apparently some factor other than Al toxicity which caused less than optimum dry matter accumulation in the pots without H-peat.

Figure 6 shows the KCl-exchangeable Al in the clay-sand mixtures as a function of soil pH and helps explain the yield curves shown in Fig. 5. For

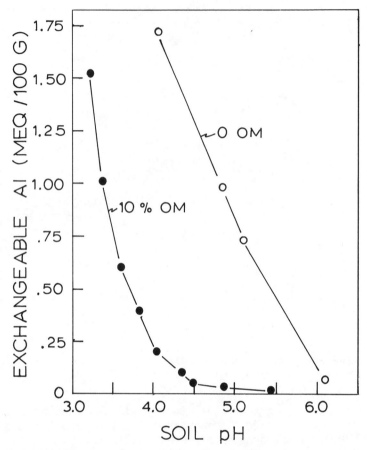

Fig. 6. The exchangeable Al content as a function of the soil pH for a sand + Al-montmorillonite mixture (0 OM) and a sand + Al-montmorillonite + H-peat mixture (10% OM).

the clay-sand mixture without organic matter, the exchangeable Al was
not reduced below the apparent toxic level until the soil was limed to
about pH 5.0 as seen by the break in the growth curve (Fig. 5). But in the
pots which had peat added, there was very little exchangeable Al at pH
4.2. In laboratory equilibration studies with peat alone, adsorption of Al
was also very high at pH 4.2, and the estimated OH/Al ratio was about 2
(Fig. 2 and 3). Apparently hydroxy-aluminum adsorbed by peat at this
pH is not very exchangeable, resulting in very little Al in solution and
good plant growth.

     To demonstrate the effect of organic matter in natural soil rather
than sand-clay mixtures, another greenhouse experiment was conducted
in which various amounts of peat were added to Zanesville subsoil (Typic
Fragiudult), an acid fragipan soil from western Kentucky. Sufficient H-
peat was added to the soil to make it 0, 2.5, 5.0, or 7.5% peat by weight,
and lime was added at seven different rates. The same barley variety
(Kearney) was grown as in the previous experiment for 6 weeks in the
greenhouse.

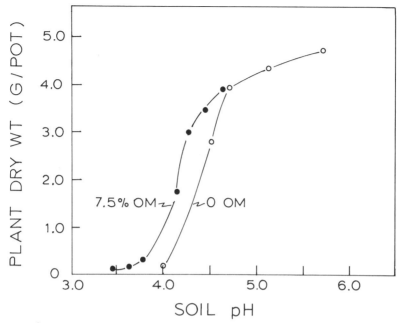

Fig. 7. Plant top dry weight as a function of soil pH for Zanesville subsoil (0 OM) and Zanesville subsoil + H-peat (7.5 % OM).

The dry matter accumulation as a function of soil pH is shown in Fig.
7 for 0 and 7.5% organic matter. Plant growth tended to be better at any
given pH with increasing organic matter levels although the differences
were not as striking as in the clay-sand mixtures. The results for 2.5 and
5.0% organic matter were intermediate between those shown in Fig. 7
but were omitted for clarity.

The relationship between soil pH and the KCl-exchangeable Al was
just as it was in laboratory equilibration studies and in the previous greenhouse experiment; the exchangeable Al was decreased at any given pH
with the addition of organic matter. This is best shown when the exchangeable Al is plotted as a function of organic matter content at a given
pH (Fig. 8), determined from the relationship between soil pH and exchangeable Al for each organic matter level (not shown).

Some plant roots from this study were excavated and washed. Roots
of plants which obviously suffered from Al toxicity were short, club-like
in shape, and coffee-brown in color. Roots growing in soil with no peat
added required a soil pH of 4.5 before roots began to show improvement
and a pH of 4.75 before roots appeared "healthy." On the other hand,
roots growing in soil with peat added appeared healthy at pH values as
low as 4.2. Even the lowest rate of peat addition apparently helped to
maintain a significantly lower solution concentration of Al over a pH
range of 4.2 to 4.8.

The differences in top growth were not very large in this study
probably because plant top growth was diminished by manganese toxicity. The Zanesville soil has a high available Mn content, and Mn concen-

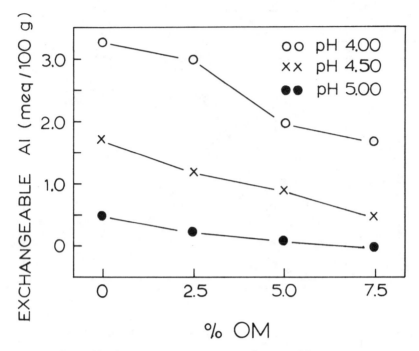

Fig. 8. Exchangeable Al content at a constant pH as a function of the organic matter content.

trations in the barley plants were as high as 800 ppm. Many plants also had visual symptoms of Mn toxicity. Therefore, although additions of organic matter decreased the amount of exchangeable Al in the soil, there was no effect on Mn concentration and thus little beneficial effect of organic matter additions on plant top growth.

Results from these experiments help explain results like those of Evans and Kamprath (1970) who found that plants grew satisfactorily on acid organic soils at a much lower pH than on mineral soils. They observed good growth of corn and soybeans at a pH as low as 4.2 on organic soils. Clark and Nichol (1966) also found that although the pH of organic soils might be quite low, very little or no lime was needed as predicted by the relationship pAl + 3pOH.

These results also help explain how corn is able to grow and yield satisfactorily in no-tillage cropping systems when the surface pH becomes very low (Blevins et al., 1977). Apparently, increases in organic matter serve to remove aluminum from the soil solution. This may also be a benefit of addition of sewage sludge or animal manures to acid soils.

SUMMARY AND CONCLUSIONS

In summary, we can compare what is known about Al-clays with what is apparent with Al-organic matter. 1) Aluminum tends to lower the acid strength of both clays and organic matter. 2) Adsorbed Al hydrolyzes in both clays and organic matter creating H^+ ions and buffering the system against increases in pH. 3) In clays, the H^+ ions produced by Al hydrolysis decompose the clay with time, reducing the CEC and releasing more Al. In organic matter the H^+ ions produced by Al hydrolysis lower the CEC by reducing the ionization of carboxyl groups. 4) In clays hydroxylated Al $[Al(OH)_x$ where $0 < x < 3]$ is non-exchangeable with neutral salts and thus blocks exchange sites unless the pH is raised to precipitate $Al(OH)_3$. Likewise, in organic matter $Al(OH)_x$ is not very exchangeable and blocks sites unless the pH is raised to form $Al(OH)_3$. 5) In clays, exchangeable Al^{3+} will maintain enough solution Al at pH < 5 to cause severe Al toxicity in plants. However, in organic matter there is apparently very little exchangeable Al^{3+} present over the normal pH range of most soils, and toxicity is severe only at very low pH (< 4.0).

With respect to Al toxicity, results from our greenhouse experiments show that increasing soil organic matter (as H-peat) tended to lower the exchangeable Al content at any given soil pH and decrease the soil pH at which Al toxicity occurred, resulting in better plant growth at any given pH with increasing organic matter levels. In addition to being of fundamental importance, these results are important in several special cases in agriculture. 1) The results help explain why plants are often able to grow and yield satisfactorily at lower pH values in organic soils than in mineral soils. 2) Under no-tillage management systems on acid soils, yields so far are less sensitive to decreases in pH. We surmise this is because the organic matter content at the soil surface tends to increase, thereby helping to alleviate Al toxicity problems. 3) Organic matter additions, such as animal manures or sewage sludge, to acid tropical soils or drastically disturbed lands which are strongly acid may help reduce Al toxicity problems.

These results seem to justify greater emphasis on the role of organic matter in soil acidity and as a cation exchanger in the soil. So far, the importance of organic matter to Al equilibria in acid soils has received only scant attention compared to reactions of Al with clay minerals. This paucity of information is probably due to the complexity of the chemical nature and behavior of soil organic matter. The gap between our knowledge of Al-clay mineral reactions and Al-organic matter reactions presents a real challenge which deserves attention.

LITERATURE CITED

1. Allison, F. E. 1973. Soil organic matter and its role in crop production. Elsevier Sci. Publ. Co., New York. p. 139–161.

2. Bartlett, R. J., and D. C. Riego. 1972. Effect of chelation on the toxicity of aluminum. Plant Soil 37:419–423.

3. Bhumbla, D. R., and E. O. McLean. 1965. Aluminum in soils: VI. Changes in pH-dependent acidity, cation exchange capacity, and extractable aluminum with additions of lime to acid surface soils. Soil Sci. Soc. Am. Proc. 29:370–374.

4. Blevins, R. L., G. W. Thomas, and P. L. Cornelius. 1977. Influence of no-tillage and nitrogen fertilization on certain soil properties after 5 years of continuous corn. Agron. J. 69:383–386.

5. Broadbent, F. E., and G. R. Bradford. 1952. Cation exchange in organic matter. Soil Sci. 74:447–457.

6. Chernov, V. A. 1947. The nature of soil acidity. (English translation furnished by Hans Jenny, translator unknown.) Soil Sci. Soc. Am., Madison, Wis.

7. Clark, J. S., and W. E. Nichol. 1966. The lime potential-percent base saturation relations of acid surface horizons of mineral and organic soils. Can. J. Soil Sci. 46:281–285.

8. Coleman, N. T., and G. W. Thomas. 1964. Buffer curves of acid clays as affected by the presence of ferric iron and aluminum. Soil Sci. Am. Proc. 28:187–190.

9. ―――, and ―――. 1967. The basic chemistry of soil acidity. In R. W. Pearson and F. Adams (ed.) Soil acidity and liming. Agronomy 12:1–41.

10. Evans, C. E., and E. J. Kamprath. 1970. Lime response as related to percent Al saturation, solution Al, and organic matter content. Soil Sci. Soc. Am. Proc. 34:893–896.

11. Flaig, W., H. Beutelspacher, and E. Rietz. 1975. Chemical composition and physical properties of humic substances. p. 1–211. In John E. Gieseking (ed.) Soil components Vol. I. Organic components. Springer-Verlag, New York.

12. Gillam, W. S. 1940. Study on the chemical nature of humic acid. Soil Sci. 49:433–453.

13. Griffith, S. M., and M. Schnitzer. 1975. The isolation and characterization of stable metal-organic complexes from tropical volcanic soils. Soil Sci. 120:126–131.

14. Harward, M. E., and N. T. Coleman. 1954. Some properties of H and Al clays and exchange resins. Soil Sci. 78:181–188.

15. Hendricks, S. B., and W. H. Fry. 1930. The results of X-ray and mineralogical examination of soil colloids. Soil Sci. 29:457–476.

16. Hester, J. B. 1935. The amphoteric nature of three coastal plain soils: I. In relation to plant growth. Soil Sci. 39:237–245.

17. Hoyt, P. B., and R. C. Turner. 1975. Effects of organic materials added to very acid soils on pH, aluminum, exchangeable NH_4, and crop yields. Soil Sci. 119:227–237.

18. Johnson, S. W. 1859. On some points of agricultural science. Am. J. Sci. Arts Ser. 2, 28: 71–85.

19. Juo, A. S. R., and E. J. Kamprath. 1979. Copper chloride as an extractant for estimating the potentially reactive aluminum pool in acid soils. Soil Sci. Soc. Am. J. 43:35–38.

20. Kelley, W. P., W. H. Dore, and S. M. Brown. 1931. The nature of the base exchange material of bentonite, soils, and zeolites, as revealed by chemical investigation and X-ray analysis. Soil Sci. 31:25–55.

21. McGeorge, W. T. 1931. Organic compounds associated with base exchange reactions in soils. Arizona Agric. Exp. Stn. Tech. Bull. 31.

22. McLean, E. O., D. C. Reicosky, and C. Lakshmanan. 1965. Aluminum in soils: VII. Interrelationships of organic matter, liming, and extractable aluminum with "permanent charge" (KCl) and pH-dependent cation-exchange capacity of surface soils. Soil Sci. Soc. Am. Proc. 29:374–378.

23. Martin, A. E., and R. Reeve. 1958. Chemical studies of podzolic illuvial horizons. III. Titration curves of organic matter suspensions. J. Soil Sci. 9:89–100.

24. Mattson, S., and J. B. Hester. 1933. The laws of soil colloidal behavior: XII. The amphoteric nature of soils in relation to aluminum toxicity. Soil Sci. 36:229–244.

25. Posner, A. M. 1966. The humic acid extracted by various reagents from a soil. Part I. Yield, inorganic components, and titration curves. J. Soil Sci. 17:65–78.

26. Reid, D. A. 1976. Screening barley for aluminum tolerance. In M. J. Wright (ed.) Plant adaptation to mineral stress in problem soils. Cornell Univ., Ithaca, N.Y.

27. Schnitzer, M., and U. C. Gupta. 1965. Determination of acidity in soil organic matter. Soil Sci. Soc. Am. Proc. 29:274–277.

28. ―――, and S. U. Kahn. 1972. Humic substances in the environment. Marcel Dekker, Inc., N.Y.

29. ————, and S. I. M. Skinner. 1963a. Organo-metallic interactions in soils: 1. Reactions between a number of metal ions and the organic matter of a podzol Bh horizon. Soil Sci. 96:86–93.

30. ————, and ————. 1963b. Organo-metallic interactions in soils: 2. Reactions between different forms of iron and aluminum and the organic matter of a podzol Bh horizon. Soil Sci. 96:181–186.

31. ————, and ————. 1964. Organo-metallic interactions in soils: 3. Properties of iron- and aluminum-organic matter complexes, prepared in the laboratory and extracted from a soil. Soil Sci. 98:197–203.

32. Stevenson, F. J. 1976. Stability constants of Cu^{2+}, Pb^{2+}, and Cd^{2+} complexes with humic acids. Soil Sci. Soc. Am. J. 40:665–672.

33. Thomas, G. W. 1975. The relationship between organic matter content and exchangeable aluminum in acid soil. Soil Sci. Soc. Am. Proc. 39:591.

34. Way, J. T. 1850. On the power of soils to absorb manure. J. R. Agric. Soc. Engl. 11: 313–379.

QUESTIONS AND ANSWERS

Q. You were plotting the yield versus the pH at various organic matter contents; however, by adding organic matter, hydrogen saturated especially, you are lowering the pH. I was wondering what your plots would look like if you plotted yields versus lime rate. My question is, given an amount of lime added, would you get just as good, or better, response with additions of organic matter?

A. What you are implying is correct. Given a constant lime rate, the amount of exchangeable aluminum would actually increase with additions of peat because the pH would decrease with additions of hydrogen saturated organic matter.

Q. In a practical sense, a very acid peat, such as you have used, would not be a reasonable amendment to add to the soil.

A. That is true, but peat does not usually occur naturally as hydrogen saturated. But, to be fair in comparing rates of organic matter additions, we used hydrogen-peat because we did not want to add additional calcium or some other nutrient in the peat. The peat we used was from Michigan and was for the most part calcium-saturated originally. We hydrogen-saturated it by leaching it with 2 N HCl.

Comment

I thought that this was an interesting paper. We have been looking at this for some time. Lowering aluminum ion activity is not the only important effect that additions of organic matter may have. Low molecular weight organic complexes may also form which may be more soluble and mobile, at pH 6 for instance, than inorganic forms. These may be far more soluble and mobile, and therefore they may have harmful effects on plants because they may move into the rhizosphere. In addition, regardless of molecular weight, organic aluminum is a great adsorber of phosphorus, and the effects are not necessarily cancelled out by liming. Phosphorus may still be adsorbed at pH 6. Neutral salt extractable or so-called exchangeable aluminum does not characterize this kind of aluminum at all and is not adequate for determining lime requirement. We have found that pH 4.8 ammonium acetate does characterize this aluminum quite well, and as a result, we have modified our soil test in Vermont. We now use the aluminum extracted by pH 4.8 ammonium acetate for lime requirement and also to modify the phosphorus recommendation.

Comment

I would like to make a comment on Hargrove and Thomas' paper. We have some data from field studies where we applied organic matter in the form of sewage sludge or chicken manure to strip mine areas. We find the very same relationship that they have talked about. Where there are high aluminum levels, the organic matter tends to overcome that, and where there are high manganese levels, it does not.

Q. I think much of what you have discussed is really dependent upon how you measure pH. I may have missed it, but did you mention how you measured pH?

A. No, I did not mention it. For the soil samples from the greenhouse experiments we used a 2:1 water to soil ratio and measured the pH using a glass electrode. However, for the equilibration experiments, when Zanesville soil was equilibrated with peat and various amounts of base, enough KCl was also added to make the solution 0.1 N KCl. So the pH of those solutions was measured in 0.1 N KCl, but the pH of the greenhouse soils was measured in water.

CHAPTER 9

Langmuir Equation and Alternate Methods of Studying "Adsorption" Reactions in Soils[1]

ROBERT D. HARTER AND GORDON SMITH[2]

The Langmuir equation has become popular as a means of describing solid-solution reactions in soils. Its popularity stems, in part from two equation coefficients, one of which provides an "adsorption maximum,"

---

[1] Scientific contribution no. 993 from the New Hampshire Agric. Exp. Stn.
[2] Associate professor of soil chemistry and former research soil chemist.

the other being related to bonding energy. As use of the equation expands, however, problems are being increasingly encountered. When development and modification of the equation is examined carefully, it becomes apparent that the equation is frequently applied incorrectly and that it may be inappropriate for describing all but a few retention reactions in soils. This is particularly true if some measure of bonding energy is desired. Most soil retention reactions are the result of several individual reactions and the availability of an "adsorbed" species is, then, dependent upon the bonding energy of each reaction. A single integrated bonding energy for the total reaction has only limited importance. Since the Langmuir Equation is based on kinetic theory, it is suggested that soil chemists consider the study of kinetics directly. Such studies have several advantages such as the ability to differentiate between different reactions occurring during the process of "adsorption," an ability to calculate reaction energies for each reaction, and an ability to obtain thermodynamic functions.

INTRODUCTION

The Langmuir equation and its potential for use in studying P adsorption reactions was brought to the attention of soil chemists during the late '50's (Fried and Shapiro, 1956; Olsen and Watanabe, 1957). A paper by Olsen and Watanabe (1957) was particularly noteworthy in that they provided an explanation of the theory behind the equation's use. This paper, in fact, is commonly cited by those who have since used the Langmuir approach. The Langmuir equation is attractive in that it provides a theoretical adsorption maximum and a coefficient which is theoretically related to bonding energy. Langmuir (1918) developed the equation on the basis of kinetic theory, but it can also·be derived using statistical mechanics (Moelwyn-Hughes, 1961; Adamson, 1967; Sposito, 1979). These facts, along with ease of application, were major factors in the technique gaining acceptability among soil chemists. The Langmuir equation is currently a common approach for studying adsorption reactions of all types.

The road to common usage has not, however, been without some doubt and apprehension expressed by those using (or rejecting) the equation. For example, Hashimoto et al. (1969) indicated that the assumption that phosphate adsorption by soils can be described by the Langmuir equation may be too optimistic. Griffin and Jurinak (1974) rejected the Langmuir equation as a means of obtaining quantitative information concerning energy of adsorption, preferring to use a kinetic approach for this purpose. Hus and Rennie (1962a, b) pointed out that a serious limitation in use of the Langmuir equation is that ions in solution are not independent in their action, whereas Langmuir (1918) had to make this assumption in his development of the equation. Hsu and Rennie (1962b) further indicated that conformation of experimental data to the Langmuir equation may not imply an adsorption reaction. Stumm and Morgan (1970) go one step further, stating that although experimental adsorption isotherms may be satisfactorily described by the Langmuir equation, this does not necessarily imply that the conditions forming the basis of the theoretical Langmuir Model are fulfilled. Misgivings regarding use of the equation reached a peak in 1977 with the publication of three papers by Veith and Sposito, Griffin and Au, and Harter and Baker. These authors called attention to certain problems experienced in use of the equation, some inappropriate use and the suggested methods to circumvent the problems.

On the other hand, a large number of soil scientists have used the equation with apparent success, obtaining adsorption maxima which realistically approximate actual adsorption and coefficients which appear to be proportional to expected bonding energy. Given the divergence of opinion regarding the utility of the Langmuir equation, this seems to be an appropriate time to carefully examine the development, modification, and current use of the equation.

THE LANGMUIR ADSORPTION EQUATION

Development of the Equation

The adsorption equation most frequently used by soil scientists is an adaption of the simple form developed for explaining gaseous adsorption on a planar surface (Langmuir, 1918). It is of the form

$$n = \frac{MbP}{1 + bP} \qquad [1]$$

Where n = the amount of gas adsorbed per unit area, P = the equilibrium gas pressure, M = the adsorption capacity, and b is a coefficient related to bonding energy. For adsorption form solutions, P has been replaced by C, or concentration in equilibrium solution. The equation can be derived from kinetic theory using the simple reaction

$$A + S \rightleftharpoons AS \qquad [2]$$

Where A is some gaseous material and S is a planar surface capable of adsorbing the gas. If θ is defined as the fraction of the surface covered, i.e.

$$\theta = \frac{n}{M} \qquad [3]$$

then the rate of gas condensation (adsorption) on the surface will be $k_aP(1 - \theta)$, where k_a is the rate coefficient in the forward direction, and the rate of evaporation (desorption) from the surface will be $k_d\theta$, where k_d is the reverse rate coefficient. At equilibrium, the rate of condensation must equal the rate of evaporation, so

$$k_d\theta = k_aP(1 - \theta) \qquad [4]$$

rearranging, this becomes

$$\theta = \frac{k_a P}{k_d + k_a P} \qquad [5]$$

and if we define

$$\frac{k_a}{k_d} = b \qquad [6]$$

Eq. 5 becomes

$$\theta = \frac{bP}{1 + bP} \qquad [7]$$

which, when combined with Eq. 3, becomes the familiar form of the Langmuir equation (Eq. 1). Thus, b in the Langmuir equation, being the ratio of the rate coefficients, is proportional to bonding energy, but should not be considered a "bonding coefficient." It is, in fact, the equilibrium constant of the reaction, since this constant is defined as the ratio of the forward to reverse rate coefficients. Therefore, free energy change, ΔG, can actually be calculated, since (Moore, 1972)

$$b = K_{eq} = \exp\left(-\Delta G / RT\right) \qquad [8]$$

where

K_{eq} is the equilibrium constant, R is the gas constant, and T is absolute temperature.

In developing Eq. 1, several assumptions were made (Langmuir, 1918):

a) the planar surfaces have a fixed number of only one kind of elementary space
b) each space is able to hold only one adsorbed molecule
c) the surface is covered with a monolayer only
d) the adsorption reaction is reversible
e) adsorbed molecules are not free to move laterally on the surface
f) adsorption energy is the same for all sites and is not dependent upon surface coverage
g) there is no interaction between adsorbate molecules
h) the probability of a molecule condensing on an unoccupied site or dissociating from an occupied site is not affected by coverage of adjoining sites.

It has been shown, however, (Adamson, 1967; Moore, 1972), that many of these assumptions are not valid, even in the situation for which the equation was derived: gaseous adsorption by planar surfaces. It is even less likely that the conditions will be met when adsorption takes place at charged surfaces in a solution. For example, even in a fairly "simple" clay-water system, we know that more than one type of adsorption site exists, that the surface can be covered by more than one layer (e.g., Stern and diffuse double layers), that even exchange reactions have an irreversible component, that adsorbed molecules can move laterally on the surface, that there is interaction between adsorbate molecules, and that adsorption energy as well as probability of adsorption or desorption is dependent upon surface coverage. Successful use of the Langmuir equation in soils research is, then, not a function of assumptions being obeyed, but is because it happens to be a good empirical equation!

Use of the Equation

THE LANGMUIR PLOT

In common use, the Langmuir equation has been rearranged to the linear form

$$\frac{C}{n} = \frac{1}{Mb} + \frac{C}{M} \qquad [9]$$

where gas pressure, P, has now been replaced by concentration in solution, C. This has the advantage that, when C/n is plotted as a function of C, the slope will be the reciprocal of the adsorption capacity, M, and the intercept will be $1/Mb$. While this treatment may yield the theoretical linear relationship, when a wide range in solution concentration is used, a curvilinear relationship is frequently reported. This anomaly has been attributed to more than one type of elementary space on the surface or the neglecting of desorbed species.

MULTIPLE ENERGY SITES

Langmuir (1918) did not restrict his development only to the simple case expressed in Eq. 1. For surfaces with more than one type of elementary space of different adsorption energies, he indicated the equation should be written in the form

$$n = \frac{M_1 b_1 P}{1 + b_1 P} + \frac{M_2 b_2 P}{1 + b_2 P} + \cdots \frac{M_n b_n P}{1 + b_n P} \qquad [10]$$

where subscripts 1 through n indicate different discrete energy adsorption sites. On this basis, various authors have divided the curvilinear "Langmuir Plots" (Fig. 1) into from two (e.g., Schuman, 1975; Rajan and Fox, 1975; Wada and Abd-Elfattah, 1978) to as many as six (Ballaux and Peaslee, 1975) separate linear portions. This division is based on the assumption that each linear portion arises from a different type of elementary adsorption site on the soil and that adsorption capacities and coefficients for each type of site can be calculated. Rather than calculating the constants from the linear portions of the curve, Holford and Mattingly (1976) have calculated them directly from Eq. 10.

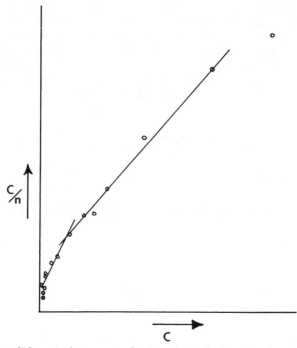

Fig. 1. "Typical" linearized Langmuir plot (Eq. 9) of soil adsorption showing curvature of plot and division into two "linear" segments. Data is for zinc adsorption by Groveton A_p horizon.

This separation of "adsorption" surfaces may still be an overly simplistic view of adsorption phenomena in soils. It is probable that Langmuir's (1918) third case, adsorption on amorphous surfaces, is the most correct view of soil adsorption. In this case, the fraction of surface covered is given by an integral expression of the form

$$\theta = \int_0^1 \frac{bP}{1 + bP} d\beta \tag{11}$$

which denotes a division of the elementary sites into an infinite number of energy levels, β being the fraction of the total number of adsorption sites that are in each energy level. If this model is correct, the curvilinear Langmuir plot (Fig. 1) may be symptomatic of something more complex than has generally been assumed, and calculation of either single or multiple adsorption capacities and coefficients, therefore, may be an exercise in futility. The values obtained in such a case would be entirely dependent upon the range of pressures (concentrations) over which data were collected and/or calculations were made. Unfortunately Eq. 11 would not be simple to evaluate for single integrated values of the equation constants.

LATERAL INTERACTION

Another potential source of curvature in the Langmuir plot could be an incorrect assumption of no lateral interaction of the adsorbate on the surface. The Langmuir equation can be modified to account for such interaction by including a lateral interaction energy term into the adsorption coefficient (Adamson, 1967). The interaction corrected adsorption coefficient (b') then becomes

$$b' = b \exp (ZE_i\theta/RT) \qquad [12]$$

where Z is the number of neighbors for each site, E_i is the lateral interaction energy, R is the gas constant, and T is absolute temperature. At low equilibrium pressure (concentration), the exponent term will be small and the resulting adsorption isotherm will differ little from those obtained if no interactions occurred. As pressure increases, however, surface coverage and lateral interaction will increase, steepening the adsorption isotherm and causing the Langmuir plot to curve. This will result in overestimation of the adsorption maximum.

USE OF EQUATION FOR SOLID-SOLUTION REACTIONS

In assuming adsorption from solution is analogous to adsorption of gases by surfaces, certain problems arise. First, whereas in gaseous adsorption the surface is reacting directly with the "matrix", in adsorption from solution the surface is reacting with a substance dissolved in the matrix. Therefore, further assumptions must be made if the Langmuir equation is to be used for these systems
   a) there is no specific adsorption of the solvent by the surface.
   b) there is no interaction between solvent and solute (Henry's Law obeyed).
   c) adsorption sites are empty "holes" having an activity coefficient of one and having zero entropy and adsorption energy (Adamson, 1967).
The first assumption is not too problematic for aqueous solutions; although water is adsorbed by surfaces, it will seldom compete with the solute for adsorption sites. The second assumption requires that all adsorption be done from dilute solutions so the activity coefficient is approximately one. In many cases, this may be undesirable. Furthermore, for electrolytes in solution, the Debye-Huckel theory states that ideal solution behavior is unattainable, even at dilute concentrations. Therefore, one must assume that electrolyte ionic interactions remain constant (Harter and Baker, 1977).

The third condition will seldom (if ever) be attained in solution and is impossible for any type of electrolytic reaction (e.g., exchange). Thus, it is more appropriate for most soil adsorption work to use a form of the equation derived to explain adsorption of one gas in a mixture of gases (Adamson, 1967)

$$n_1 = \frac{M_1 b_1 P_1}{1 + b_1 P_1 + b_2 P_2 + \ldots b_n P_n} \qquad [13]$$

where the subscripts denote different gases. In soil adsorption from solution, P_2, P_3, . . ., P_n become concentrations of desorbed species in solution. Harter and Baker (1977) have argued that ignoring these desorbed species has been another cause of nonlinearity in the Langmuir plot. A method by which Eq. 13 can be evaluated for the various constants was proposed by these authors. For the most precise evaluation of adsorption, then, one should probably combine Eq. 10 (or 11), 12, and 13. However, evaluation would obviously become a "horror," and even this equation would not be completely accurate, since Eq. 13 assumes exchange of single ions or molecules on single sites. Thus, Eq. 13 is strictly applicable only for exchange of monovalent ions. Exchange of polyvalent and particularly exchange of ions having different charges introduces additional complexities to the relationship.

MECHANISMS OF RETENTION

If retention of adsorbate by surfaces is via a simple exchange reaction, use of the Langmuir equation to describe the process is reasonable. However, a major problem in using this equation to describe soil adsorption is that the adsorbate may be retained by a combination of mechanisms. The Langmuir equation was developed assuming adsorption is via a simple physio-chemical retention by the surface. As indicated, this model can be modified for multiple elementary sites, for ion exchange, for lateral interaction at the surface, etc., but the basic model remains the same. "Adsorption" by soils is a much more complex phenomena (therefore, the quotes in the title of this paper). Some ion exchange certainly occurs, but to assume this accounts for all retention is shortsighted. We know that many other reactions also account for ion retention by soils. Such reactions include precipitation-crystalization, structural substitutions, chelation with organic matter, etc. While these reactions bring about a reduction in solution concentration, they cannot be considered adsorption reactions. Furthermore, attempts to apply theories developed to quantify adsorption processes to quantification of these nonadsorption processes are inappropriate. The indicated complexity of retention reactions is best illustrated by kinetics research. For example, an initial second order reaction followed by a first order reaction such as found by Griffen and Jurinak (1974) for phosphate adsorption onto calcite has frequently been reported. We have found a similar progression for Cu retention by soil (Fig. 2), with the exception that the second order reaction is preceded by an instantaneous reaction that is completed in less than 15 sec. In addition, based on Ca in solution and hydroxide demand, following the first order reaction there may be some reordering of the

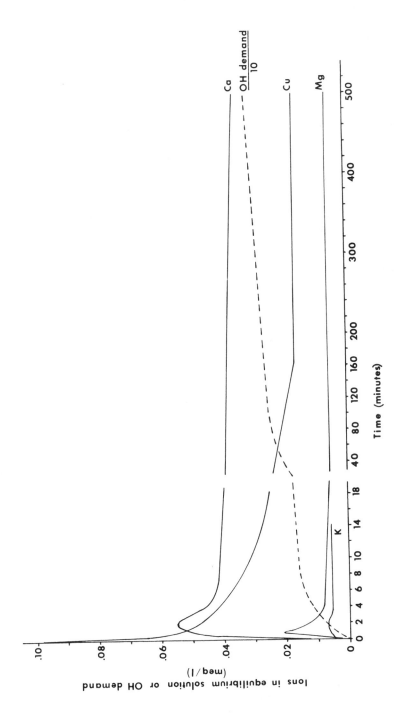

Fig. 2. Kinetics of copper reaction with Christiana A_p horizon at 25 C.

ions. These multiple reactions cannot be satisfactorily characterized by a single equation, since each component of the reaction will have a specific bonding energy and adsorption maximum. Whether, for example, a component having a high bonding energy has a high or low adsorption capacity is of significance when studying potential availability of ions. Only by use of kinetic reactions can the various reactions be delineated in this manner.

It seems, therefore, that the time has come to reassess the use of the Langmuir equation, perhaps reordering our thinking regarding what can be obtained by its use. The Langmuir equation might be useful, for example, in the characterization of the initial instantaneous reaction (Fig. 2) that occurs too rapidly to follow kineticly, but not for the entire reaction. (Since the Langmuir equation is based on equilibrium conditions, equilibrium concentrations upon completion of the theoretical reaction would have to be extrapolated.) As indicated earlier, it is a good empirical equation which frequently describes adsorption better than the Freundlich equation. Therefore, the Langmuir equation might be useful when a researcher is simply looking for numerical relationships to describe data. Furthermore, it is often possible to calculate a realistic adsorption maximum from the equation. However, we should also realistically assess the shortcomings of the equation, particularly with regard to its ability to differentiate between different reactions occurring in the soil. Other methods of studying "adsorption" phenomena in soils may be more appropriate under certain circumstances.

ALTERNATIVES TO THE LANGMUIR EQUATION

Adsorption Isotherms

In many cases, it is not necessary to use the equation to obtain an adsorption maximum. If a sufficient breadth of solution concentration is used to obtain an isotherm such as illustrated in Fig. 3B, the maximum adsorption can be readily estimated from the plot. The solution concentrations necessary to complete such an isotherm have frequently been cited as unrealistic compared to those normally found in nature. However, isotherms constructed by incremental adsorption from a dilute solution normally are identical to those constructed in the more traditional manner of adding varying concentrations of an adsorbate. When excess retention does occur when the system is subjected to a high concentration, the excess adsorbate can be removed by a single water wash. As a result, we have incorporated a water wash into our procedures to check for excess retention. In addition, when simulating "real world" phenomena the high solution concentrations used may not be so excessive as first appears. For example, while we might think in terms of a 100 ppm (200 lb./A) P fertilizer addition, the fertilizer is not added to the whole plow layer, but may be added as a small band. If the bands are 76 cm (30 inches) apart and the P reacts with an area of 2.5 cm (1 inch) diam, the P concentration would be more like 23,000 ppm.

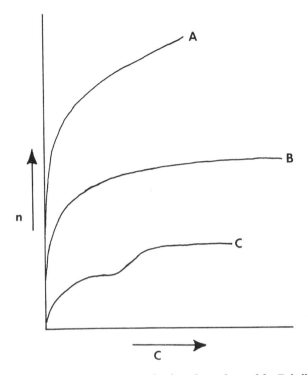

Fig. 3. Some "typical" adsorption isotherms that have been observed for Dekalb A_p horizon. A. Pb adsorption at pH 6.2, B. Pb adsorption at pH 5.3, C. Ni adsorption at pH 5.3.

When an isotherm such as Fig. 3A is obtained, it may be desirable to calculate an adsorption maximum with help of the Langmuir equation. However, even with this amount of data, it generally is possible to project the isotherm to a theoretical plateau. The adsorption maximum thus obtained is often as good or better than that calculated.

It may not be possible to duplicate Fig. 3C by using incremental addition of dilute solutions, but this type of relationship is occasionally found in adsorption studies. This cannot be handled with the Langmuir equation. Some modified form of the BET equation (Adamson, 1967) might fit this isotherm, but it would be simpler to take at least the second maximum from the plot.

Kinetic Approaches

Although, as has been indicated (Eq. 4 through 8), change in free energy can actually be calculated from Langmuir equation parameters, the problems and complexities encountered with the use of this equation cause a significant degree of uncertainty in the results. It seems reasonable, therefore, to use kinetic approaches to describe adsorption wherever possible. In addition to better understanding energetics of the reaction(s), insight into the actual "adsorption" mechanisms can be obtained.

THE ARRHENIUS EQUATION

Thermodynamic parameters can be calculated from kinetics data. As a first step, the Arrhenius equation

$$\frac{d \ln (k_r)}{dT} = \frac{Ea}{RT^2} \qquad [14]$$

(where k_r is the reaction rate constant, T is absolute temperature, R is the gas constant, and Ea is the Arrhenius activation energy) must be evaluated (Avery, 1974). Ea is, in itself, a useful value, since it indicates the size of the energy barrier that must be crossed for the reaction to proceed (Fig. 4). Equation 14 can be integrated to the form

$$k_r = A \exp (-Ea/RT) \qquad [15]$$

where A is a constant known as the frequency factor (Avery, 1974). If, then, $\ln k_r$ is plotted as a function of $1/T$, the intercept ($1/T = 0$) will give the constant, A, and the slope will be equal to $-Ea/R$. Alternately, Eq. 14 can be integrated to the form:

$$\ln \left(\frac{k_2}{k_1} \right) = \frac{Ea}{R} \left[\frac{T_2 - T_1}{T_1 T_2} \right] \qquad [16]$$

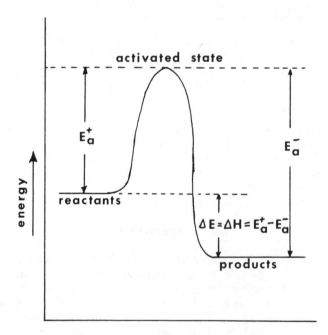

Fig. 4. Hypothetical potential energy diagram for an exothermic reaction.

It may be somewhat easier to evaluate the parameter by measuring rate constants at two temperatures and using Eq. 16 to calculate Ea. If the frequency factor is needed, then, this value of Ea can be substituted into Eq. 15 at one of the temperatures (Avery, 1974).

RATE CONSTANTS

In soil reactions, the forward and reverse reactions will normally be occurring simultaneously. Therefore, to evaluate the forward and reverse activation energies and the enthalpy (ΔH) of the reaction, it will be necessary to ascertain the forward and reverse rate constants (k_a and k_d).

Considering two first order opposing reactions,

$$A \underset{k_d}{\overset{k_a}{\rightleftharpoons}} B \qquad [17]$$

initially, the concentrations of A will be a, and B will be zero. At time, t, we will let the concentration of B be x, and A will be (a-x). The rate of reaction can be written

$$-\frac{d[A]}{dt} = \frac{dx}{dt} = k_a (a\text{-}x) - k_d x \qquad [18]$$

at equilibrium, Eq. 18 becomes

$$0 = k_a (a\text{-}x_e) - k_d (x_e) \qquad [19]$$

where x_e is the concentration of B at equilibrium. Rearranging and substituting back into Eq. 18, we get

$$\frac{dx}{dt} = k_a \left[\frac{a (x_e - x)}{x_e} \right] \qquad [20]$$

integrating,

$$k_a = \frac{x_e}{at} \ln \left[\frac{x_e}{x_e - x} \right] \qquad [21]$$

substituting this value for the forward rate constant back into Eq. 19 we find that

$$k_d = \frac{1}{t} \ln \left[\frac{x_e}{x_e - x} \right] - k_a \qquad [22]$$

obviously, these equations would be of slightly different form if other than a first order equation occurs, but the adjustment should be relatively simple.

THERMODYNAMIC PARAMETERS

The calculated values for the forward and reverse rate constants can be used to obtain the forward and reverse activation energies from which the enthalpy of reaction can be calculated from the relationship

$$\Delta H = \Delta E - \Delta n (RT) \qquad [23]$$

where Δn is the number of moles of product less the number of moles of reactant. In most solid-solution "adsorption" reactions, Δn would be zero, so ΔE will be equivalent to enthalpy change. It may be possible, however, to have organic reactions in which Δn would not be zero.

Since the equilibrium constant is the ratio of the forward to reverse rate constants (Eq. 8), kinetics can also be related to other thermodynamic quantities. For example, we can directly calculate the free energy change for the reaction by substituting into Eq. 8

$$\ln k_a - \ln k_d = \ln K_{eq} = \Delta G/RT \qquad [24]$$

Finally, entropy change for the reaction can be obtained from the familiar expression

$$\Delta G = \Delta H - T\Delta S \qquad [25]$$

Alternately, entropy of the forward and reverse reactions can be separately calculated from the Arrhenius frequency factor, since

$$A = \frac{\varkappa T}{h} \exp \left(m + \frac{\Delta S_a}{R} \right) \qquad [26]$$

where $\varkappa$ is the Boltzmann constant, h is Planck's constant, and m is the molecularity of the equation and ΔS_a is the change in entropy for the activated complex (Avery, 1974). Entropy change for the reaction will be the difference between forward and reverse entropy change.

CONCLUSIONS

Considering the problems and imprecisions that have been encountered is use of the Langmuir equation, it seems appropriate that new approaches to the study of soil adsorption phenomena be considered. We do not suggest that the Langmuir equation be abandoned entirely, only that it should not be used universally. When reaction energies are desired, alternatives such as kinetics should be considered. Accurate bonding energies should, in particular, be obtained for soils from land areas to be used for disposal of potentially toxic wastes. While the total adsorption capacity is still of interest in such areas, the ability of a soil to hold a retained ion against leaching or plant uptake is of critical importance. As indicated, the thermodynamic functions of adsorption can be calculated from kinetic studies. Such studies are not difficult, as has been shown by Zasoski and Burau (1978), and have the added benefit of providing information on bonding mechanisms. Furthermore, if more than one separable reaction is occurring during "adsorption" the thermodynamic functions can be separately calculated for each reaction.

LITERATURE CITED

1. Adamson, A. W. 1967. Physical chemistry of surfaces, 2nd ed. John Wiley & Sons, Inc., New York.
2. Avery, H. E. 1974. Basic reactions kinetics and mechanisms. The Macmillan Press, Ltd., London, England.
3. Ballaux, J. C., and D. E. Peaslee. 1975. Relationships between sorption and desorption of phosphorus in soil. Soil Sci. Soc. Am. Proc. 39:275–278.
4. Fried, Maurice, and R. E. Shapiro. 1956. Phosphate supply pattern of various soils. Soil Sci. Soc. Am. Proc. 20:471–475.
5. Griffin, R. A., and A. K. Au. 1977. Lead adsorption by montmorillonite using a competitive langmuir equation. Soil Sci. Soc. Am. J. 41:880–882.
6. ————, and J. J. Jurinak. 1974. Kinetics of the phospahte interaction with calcite. Soil Sci. Soc. Am. Proc. 38:75–79.
7. Harter, R. D., and D. E. Baker. 1977. Application and misapplications of the Langmuir equation to soil adsorption phenomena. Soil Sci. Soc. Am. J. 41:1077–1088.
8. Hashimoto, Isao, J. D. Hughes, and O. D. Philen, Jr. 1969. Reactions of triammonium pyrophosphate with soils and soil minerals. Soil Sci. Soc. Am. Proc. 33:401–405.
9. Holford, I., C. R. and G. E. G. Mattingley. 1976. A model for the behavior of labile phosphate in soil. Plant Soil. 44:219–229.
10. Hsu, Pa Ho, and D. A. Rennie. 1962a. Reactions of phosphorus in aluminum systems. I. Adsorption of phosphate by X-ray amorphous "aluminum hydroxide." Can. J. Soil Sci. 42:197–209.
11. ————, and ————. 1962b. Reactions of phosphorus in aluminum systems. II. Precipitation of phosphate by exchangeable aluminum on a cation exchange resin. Can. J. Soil Sci. 42:210–221.
12. Langmuir, Irving. 1918. The adsorption of gasses on plane surfaces of glass, mica, and platinum. J. Am. Chem. Soc. 40:1361–1382.
13. Moelwyn-Hughes, E. A. 1961. Physical chemistry, 2nd ed. Macmillan Co., New York.
14. Moore, W. J. 1972. Physical chemistry, 4th ed. Prentice-Hall, Inc., Englewood Cliffs, N.J.
15. Olson, S. R., and F. S. Watanabe. 1957. A method to determine a phosphorus adsorption maximum of soils as measured by the Langmuir isotherm. Soil Sci. Soc. Am. Proc. 21:144–149.
16. Rajan, S. S. S., and R. L. Fox. 1975. Phosphate adsorption by soils: II. Tropical acid soils. Soil Sci. Soc. Am. Proc. 39:846–851.
17. Schuman, L. M. 1975. The effect of soil properties on zinc adsorption by soils. Soil Sci. Soc. Am. Proc. 39:454–458.
18. Sposito, Garrison. 1979. Derivation of the Langmuir equation for ion exchange in soils. Soil Sci. Soc. Am. J. 43:197–198.
19. Stumm, Werner, and J. J. Morgan. 1970. Aquatic chemistry. An introduction emphasizing chemical equilibria in natural waters. Wiley-Interscience Pub. Co., New York.
20. Veith, J. A., and Garrison Sposito. 1977. On the use of the Langmuir equation in the interpretation of "adsorption" phenomena. Soil Sci. Soc. Am. J. 41:697–702.
21. Wada, Koji, and Aly Abd-Elfattah. 1978. Characterization of adsorption sites in two mineral soils. Soil Sci. Pl. Nutr. 24:417–426.
22. Zasoski, R. J., and R. G. Burau. 1978. A technique for studying the kinetics of adsorption in suspensions. Soil Sci. Soc. Am. J. 42:372–374.

QUESTIONS AND ANSWERS

Q. Is not another assumption basic to the Langmuir adsorption equation; this being that there is no change in the energy of the solid specie as might occur if the clay mineral structure were altered by the adsorption of a particular cation?

A. Yes. It is an understood assumption that the solid is "inert". Reactions leading to energy changes within the solid would be neither surfacial nor adsorption.

Q. Does the Langmuir equation depend on the assumption that the absorbing plane is in contact with a pure gas?

A. No. Eq. 13 is based on the assumption that absorption of one gas from a mixture of gases is possible. Now, Eq. 1, the simple Langmuir equation, does assume a pure gas, but it can be readily modified to account for adsorption of a single gas from a mixed gas system.

Q. Would it not be true that if you were to consider b, the equilibrium constant, you would have to deal in activities rather than concentrations?

A. That's probably true, however, I don't even want to consider b any longer, at least in terms of a relationship to energy. There are so many other problems involved in attempting to use b as a measure of energy that it should be considered nothing more than an equation coefficient and no physical significance should be associated with it. In this case, it would not matter whether concentration or activity were used. The point I am making is that the Langmuir equation should be used for nothing more than an adsorption maximum, and caution should be exercised even in this usage.

Q. Could you use the Langmuir bonding energy for relative differences between two different soils or between two different soil layers just to show relative differences on a qualitative basis, not on a quantitative basis?

A. I would be very reluctant to do so. The equation coefficient might be different for each soil or horizon, but based on my studies of the last six months or so, I really feel that b has little relationship to bonding energy. Because we have such a complex equilibrium system, if we want energies we'd better be prepared to go to kinetics to get them. An additional problem is its past history. Even if the equation coefficient were used qualitatively with no association to energy, this connection would persist in many minds. A complete break is necessary.

CHAPTER 10

Solid Phase-Solution Equilibria in Soils[1]

W. L. LINDSAY[2]

INTRODUCTION

Soils are complex chemical systems in which the dissolution and precipitation of solid phases largely control solubility relationships. How can

---

[1] Contribution from the Colorado State Univ. Exp. Stn., Fort Collins, and published as scientific series paper no. 2526.

[2] Department of Agronomy, Colorado State Univ., Fort Collins, CO 80523.

our knowledge of chemistry be used to describe the chemical reactions and solubility relationships that occur in soils?

Recent advances have been made in understanding solid phase-solution equilibria. Many specific minerals and solid phases found in soils have been identified and characterized (Dixon and Weed, 1977). Garrels and Christ (1965) have shown how the solubility relationships of various minerals found in geochemical environments can be described while Stumm and Morgan (1970) have shown similar relationships for aqueous environments. A recent selection and compilation of the standard free energies of formation for use in soil chemistry should also be helpful (Sadiq and Lindsay, 1979).

Recently, the author has completed a text entitled *Chemical Equilibria in Soils* (Lindsay, 1979). The central thrust of this development has been to express the solid phase-solution equilibria in soils in quantitative terms.

The present paper offers examples of how the solid phase-solution equilibria in soils can be expressed mathematically to give valuable insights into chemical reactions and solubility relationships. The author has found equilibrium relationships to be extremely useful in soil chemistry for developing working hypotheses and for focusing research efforts. Students also appreciate the approach because many seemingly unrelated facts and observations can be organized into diagrams where meaningful relationships can be readily seen and tested.

EQUILIBRIUM IN SOILS

Some reactions in soils attain equilibrium very rapidly while others attain it more slowly. The formation of soils from parent materials and the weathering of rocks and minerals over geological periods are examples of chemical reactions that proceed very slowly.

Equilibrium provides a valuable reference point that is useful in ascertaining the direction of any chemical reaction. For example, the precipitation of minerals can only occur under conditions of supersaturation. The dissolution of minerals can only occur under conditions of undersaturation. All chemical reactions proceed toward equilibrium, never away from it. Thus, equilibrium relationships provide the key to predict which reactions can and which cannot occur.

Too often in soil science we dwell on the complexity of soil rather than its simplicity. Exchange reactions and organic matter transformation, which are fairly complex, have attracted great attention, yet beneath these veneers lie the mineral equilibria which have ultimate control of solubility relationships. The presence of solid phases in soils greatly simplify solubility considerations. The free energy of a crystalline phase is fixed and thereby buffers the free energy of the constituent ions in solution.

It is important to understand the nature of solid phases in soils. Initial precipitates are often amorphous and have a higher free energy than crystalline phases of similar composition. Thus, solubility relationships in soils may change as solid phase transformations occur. Many such transformations take place so slowly that impatient and unknowing researchers falsely conclude that equilibrium relationships in soils are of little value.

Soils contain many different elements. Some of these participate in isomorphous substitutions that may distort crystal lattices and alter their free energy relationships slightly. Such phenomena need careful study and examination but should not destroy our confidence in dealing with soils primarily as chemical systems. It has been the author's experience that when the principles of chemistry are rigorously applied to soils, the predictions correspond amazingly well to observed behavior.

Soils are dynamic systems in which many different reactions are occurring simultaneously. These reactions involve numerous solid, solution, and gas phases. Equilibrium relationships provide the unifying principle by which each reaction can be examined and analyzed. Without such tools, soil chemistry becomes a helpless maze of empirical relationships.

SOLID PHASE REDOX RELATIONSHIPS

Reduction-oxidation (redox) is important in soils because it modifies solubility relationships and brings about many mineral transformations. Electrons are continually released to soils through the metabolic activities of living organisms. Normally these electrons combine with oxygen, and soils are maintained in a fairly well-oxidized condition. Whenever the oxygen supply to any part of a soil is restricted, however, electrons accumulate and are available to reduce other components in the soil.

Most soil scientists find theoretical redox relationships perplexing and generally avoid their use. Much of the soils literature on redox consists of merely reporting redox measurements and repeating Eh-pH diagrams from the geochemical literature. Only occasionally have theoretical redox relationships been critically used to interpret experimental data or to predict many of the fascinating redox relationships that occur in soils. Examples are the excellent review by Ponnamperuma (1972) and the elaborate documentation of redox measurements in the environment by Bass Becking et al. (1960), which have contributed significantly to our knowledge of redox relationships in soils and other natural environments.

pe vs. Eh

Much of the perplexity associated with theoretical redox relationships arises from the fact that redox is generally expressed in terms of Eh (volts or millivolts). Redox reactions expressed in this manner are difficult to combine with other chemical equilibria which are expressed by means of equilibrium constants. Sillen and Martell (1964) used the term pE, or more appropriately pe, to express redox, where pe is the negative log of the electron activity (e^-) based on the standard hydrogen half cell in which the electron activity is arbitrarily set as unity. This convention permits both redox and other chemical reactions to be combined and their equilibrium expressed by a single equilibrium constant. Although many have recognized advantages in using pe (Truesdell, 1969; Stumm and Morgan, 1970; Ponnamperuma, 1972), its use in soils along with the parameter pe + pH has only recently been more fully explored (Lindsay, 1979; Lindsay and Sadiq, 1981). The redox range of aqueous systems expressed in terms of pe + pH lies between 0 (1 atm of H_2) and 20.78 (1 atm of O_2).

pe + pH as a Redox Parameter

Whenever two oxide, hydroxide, carbonate, or silicate minerals contain a given element in more than one oxidation state, equilibrium between the two minerals fixes the redox at a constant pe + pH. For example, consider the manganese minerals MnO_2 (pyrolusite) and $MnOOH$ (manganite) in which the oxidation states are Mn(IV) and Mn(III), respectively. Equilibrium between the two minerals can be expressed as:

			log K°
β-MnO_2 (pyrolusite) + $4H^+ + 2e^-$	$\rightleftharpoons$	$Mn^{2+} + 2H_2O$	41.89
$Mn^{2+} + 2H_2O$	$\rightleftharpoons$	γ-$MnOOH$ (manganite) $+ 3H^+ + e^-$	-25.27
β-MnO_2 (pyrolusite) $+ H^+ + e^-$	$\rightleftharpoons$	γ-$MnOOH$ (manganite)	16.62 [1]

for which the equilibrium expression is

$$\frac{1}{(H^+)\,(e^-)} = 10^{16.62} \qquad [2]$$

or

$$pe + pH = 16.62 \qquad [3]$$

These relationships show that as long as both pyrolusite and manganite are present and equilibrium is maintained, the pe + pH is fixed at 16.62. If additional electrons or protons are added, Reaction 1 proceeds to the right with dissolution of pyrolusite and precipitation of manganite, but the pe + pH remains at 16.62. Similar reactions can be written for other mineral combinations in which an element is present in more than one oxidation state.

Electron Titration of Soils

An electron titration curve for a hypothetical soil was developed by Lindsay and Sadiq (1981) and is reproduced in Fig. 1. The various pe + pH poises are numbered to correspond to the designated mineral transformations indicated in the legend. For this development an arbitrary

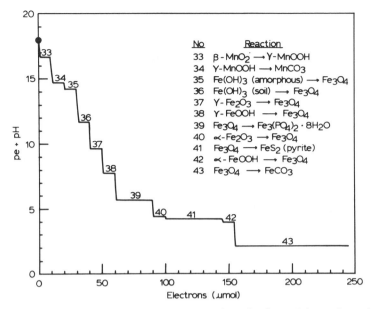

Fig. 1. Calculated electron titration curve for a hypothetical soil containing various minerals showing the various pe + pH poises during reduction (Lindsay and Sadiq, 1981).

amount of each mineral was assumed so that the various redox poises could be systematically represented. In actual soils the quantity of electrons needed for each transformation would differ depending on the amount of reducible mineral present. The pe + pH of each redox poise aids in identifying the mineral transformation occurring there.

Electron titration curves are useful tools for understanding solid phase-redox relationships in soils, but they are difficult to obtain experimentally. Near-equilibrium conditions must be maintained between the two minerals involved. This can only be achieved if the electron titration is carried out very, very slowly (Sadiq, 1977). Furthermore, the redox poises must be sufficiently separated (approximately 0.5 units) if they are to aid in identifying specific mineral transformations.

Regardless of the problems involved in carrying out electron titrations of soils, the principles involved are of fundamental importance to soil chemistry and offer many challenges for further research.

EXAMPLES OF METAL ION EQUILIBRIA IN SOILS

Metal ions often precipitate in soils. If such solid phases can be identified and their solubilities determined, useful stability diagrams can be developed. Many trace elements are present in soils in such small quantities that it is virtually impossible to separate and identify discrete mineral phases. In such cases solubility measurements can be made directly on soils, and the results can be expressed as empirical soil parameters such as Soil-Fe, Soil-Zn, Soil-Cu, etc.

Soil-metal Equilibrium Reactions

Norvell and Lindsay (1969, 1972, 1981) and Lindsay and Norvell (1969) reacted metal chelates with soils and showed how metal ion activities can be obtained and expressed by equilibrium reactions. Their results with well-oxidized soils include:

		log $K°$	
Soil-Fe + 3H$^+$ $\rightleftharpoons$ Fe^{3+}		2.70	[4]
Soil-Zn + 2H$^+$ $\rightleftharpoons$ Zn^{2+}		5.80	[5]
Soil-Cu + 2H$^+$ $\rightleftharpoons$ Cu^{2+}		2.80	[6]

These empirical equations are useful to express the pH-dependent solubility relationships often found in soils.

In contrast, cations like Ca^{2+} and Mg^{2+} are controlled by the exchange complex in acid soils because most Ca and Mg minerals are too soluble to persist below pH 7 (Lindsay, 1979). Such solubility relationships can be approximated by the equilibrium expressions:

	log $K°$	
Soil-Ca $\rightleftharpoons$ Ca^{2+}	-2.50	[7]
Soil-Mg $\rightleftharpoons$ Mg^{2+}	-3.00	[8]

in alkaline soils Ca^{2+} is generally controlled by calcite while Mg^{2+} behaves as if it were controlled by calcite-dolomite, "magnesian calcites" or other Mg minerals of similar solubility (Lindsay, 1979; Hassett and Jurinak, 1971). Such solubility relationships can be expressed as:

		log $K°$	
$CaCO_3$ (calcite) + 2H$^+$	$\rightleftharpoons$ Ca^{2+} + CO$_2$ (g) + H$_2$O	9.72	[9]
$MgCa(CO_3)_2$ (dolomite) + 2H$^+$	$\rightleftharpoons$ Mg^{2+} + CO$_2$ (g) + H$_2$O + CaCO$_3$(c)	8.70	[10]

Thus, metal ion activities in the soil solution can be related to pH (Fig. 2). Similar soil-metal reactions can be obtained for other metal ions as well. Care must be taken, however, to insure that speciation of the metal ions is taken into consideration, otherwise the soil-metal equilibrium reactions are of only limited value. Chelating agents have been used to measure metal ion activities (Norvell and Lindsay, 1981).

Solubility of Zinc Minerals in Soils

An example of how the solubility relationships of a metal ion such as zinc can be represented in soils is shown in Fig. 3 (Lindsay, 1979). As other Zn minerals are recognized and their solubilities become known,

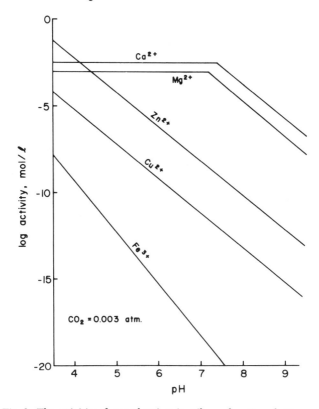

Fig. 2. The activities of several cations in soils as a function of pH.

they too can be added to this diagram. Similar diagrams can be developed for other metal ions and their associated minerals.

The more soluble Zn minerals such as hydroxides, oxides, and carbonates are unstable in soils, and if added, they will eventually dissolve. Such minerals should constitute good Zn fertilizers because they support levels of Zn that are generally adequate for plants (Lindsay, 1972). The mineral Zn_2SiO_4(willemite) is of intermediate solubility whereas Soil-Zn described in the previous section may lower Zn^{2+} activity even further. The ferrite mineral $ZnFe_2O_4$(franklinite) is very stable in soils and can control Zn^{2+} near the Soil-Zn line. The solubility of Zn^{2+} maintained by franklinite is affected by the solubility of iron. If Fe^{3+} activity is controlled between that maintained by Soil-Fe and γ-Fe_2O_3(maghemite), franklinite, and Soil-Zn would have similar solubilities.

Effect of Redox on Copper Solubility

Redox can affect the solubility and mineral transformations of metals as illustrated for Cu (Fig. 4) (Lindsay, 1979). In this diagram the activities of both Cu^{2+} and Cu^+ are plotted as a function of pe + pH in the range of zero (equilibrium with 1 atm H_2) to 20.78 (equilibrium with 1

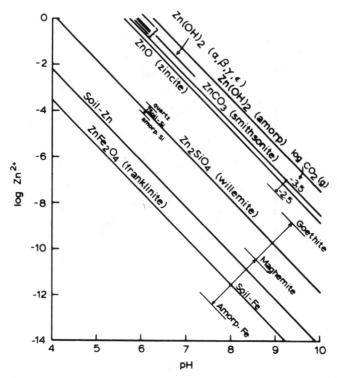

Fig. 3. The solubilities of several Zn minerals compared to Soil-Zn (Lindsay, 1979).

atm of O_2). The graph is drawn for pH 7 with shifts shown for pH 6 and 8. Above pe + pH of 14.89, Soil-Cu controls Cu^{2+} at approximately $10^{-11.20}M$. In the pe + pH range of 14.89 to 11.53 Cu solubility is controlled by $Cu_2Fe_2O_4$ (cuprous ferrite) and Soil-Fe. Between pe + pH of 11.53 and 4.73 Cu solubility is controlled by $Cu_2Fe_2O_4$ (cuprous ferrite) and Fe_3O_4 (magnetite). When redox drops below 4.73, $Cu_2S(c)$ can precipitate based on a reference level of SO_4^{2-} of $10^{-3}M$. Other Cu minerals such as CuO(tenorite), $Cu(OH)_2(c)$, Cu_2O (cuperite), CuOH(c) and Cu(c) have been shown to be unstable in soils (Lindsay, 1979). Diagrams such as Fig. 4 contain useful information that can aid in deciphering many of the complex solubility relations of Cu in soils.

Effect of Redox on Manganese Solubility

Redox greatly influences the solubility of some elements as illustrated for Mn (Fig. 5) (Sadiq, 1977; Lindsay, 1979). Under highly-oxidized conditions β-MnO_2(pyrolusite) appears as the most stable manganese mineral. Mixed valence minerals such as γ-$MnO_{1.9}$(nsutite) and δ-$MnO_{1.8}$ (birnessite) are slightly more soluble. As redox lowers to 16.62, γ-MnOOH(manganite) becomes the stable phase (Eq. 1 through 3). The

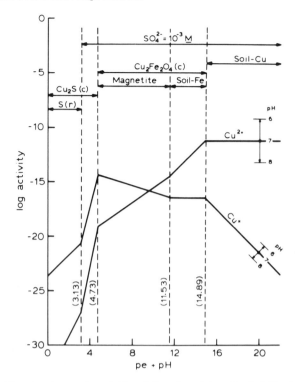

Fig. 4. The effect of redox on the stability of Cu minerals in soils of pH 7 with shifts shown for pH 6 and 8 (Lindsay, 1979).

fact that pyrolusite, nsutite, birnessite, and manganite have similar solubilities at this redox, accounts for the fact that Mn often forms unidentifiable, mixed valence precipitates with changing redox (Zordan and Hepler, 1968; Ponnamperuma et al., 1969).

As pe + pH drops below 16.62, γ-MnOOH(manganite) becomes the stable phase (Fig. 5). Later $MnCO_3$(rhodochrosite) becomes stable depending upon $CO_2(g)$. When it is high, less reduced conditions are necessary for rhodochrosite to form. The mineral Mn_2O_3(bixbyite) is always metastable to MnOOH(manganite). Only at $CO_2(g)$ levels considerably below that of the atmosphere ($10^{-3.52}$ atm) can Mn_3O_4(hausmannite) form. Furthermore, pe + pH would have to drop below 12.78 (at pH 7.0) and Mn^{2+} would have to exceed 0.1 M for hausmannite to form. It would also be virtually impossible for $Mn(OH)_2$(pyrochroite) to form in soils.

The manganese solubility diagram in Fig. 5 was drawn for pH 7, but shifts for other pH values can be readily obtained because all lines shift up or down 2 log units for each unit change in pH as noted in the upper right corner of the diagram. Experimental findings with manganese (Gotoh and Patrick, 1972) roughly correspond to these predictions. As other manganese minerals are recognized as being important in soils (McKenzie, 1977) and their solubilities become known, they too can be included.

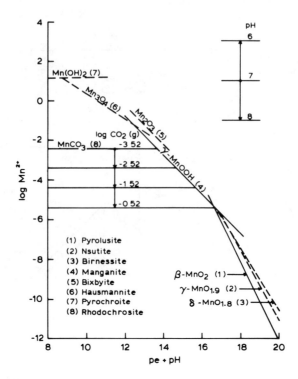

Fig. 5. The effect of redox and $CO_2(g)$ on the solubility and stability of Mn minerals at pH 7 (Lindsay, 1979).

Speciation of Iron in Solution

Most metal ions hydrolyze in aqueous solutions and also form various ion complexes. To account for total solubility of a given metal in solution or to relate total solubility to the activity of a specific ion, speciation of that element must be known. The speciation of Fe complexes and their activities in equilibrium with Soil-Fe are shown in Fig. 6.

Ranges of anion activities were selected to correspond to those expected in soils. Only the hydrolysis species of Fe^{3+} are significant in the pH range of most soils. The other complexes shown here do not contribute significantly to total Fe in solution.

Iron solubility is also affected by changes in redox. A similar diagram can be constructed for Fe^{2+} and its hydrolysis and complex ions. It has been shown that Fe(II) and its solution complexes can only exceed those of Fe(III) as the pe + pH drops below 12.0 (Lindsay, 1979). Such relationships help to explain why soils must be reduced below this redox level before the solubility and availability of iron is increased.

Not represented in Fig. 6 are the natural organic matter-iron complexes that are present in soil solution. Such complexes raise the total level of Fe in solution and contribute to its mobility and availability to plants

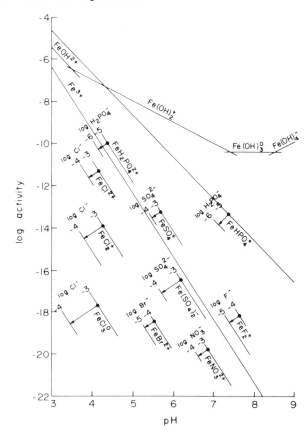

Fig. 6. Complexes of Fe(III) in equilibrium with Soil-Fe compared to the hydrolysis species of Fe^{3+}.

(Lindsay, 1974). These complexes need further study and characterization. Soil suspensions appropriately shaken in the presence and absence of C black have been used to measure the extent of organic complexing as most organic complexes can be adsorbed onto charcoal black (Moreno et al., 1960).

Aluminosilicate Equilibria

Aluminosilicates comprise a significant fraction of soils. Most primary minerals were formed at temperatures and pressures much different than those that exist at the earth's surface today. Most primary minerals slowly dissolve as the constituents that are released recombine to form more stable secondary soil minerals.

The solid phase-solution equilibria of several important aluminosilicate minerals can be plotted as a function of log $H_4SiO_4^0$ (Fig. 7) (Lindsay, 1979). Similar relationships have been reported (Kittrick, 1971; Rai

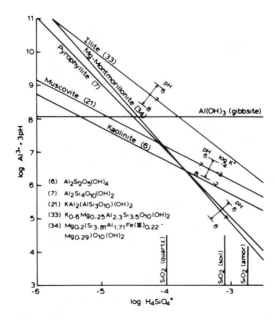

Fig. 7. The stability relationships of several aluminosilicates in equilibrium with Soil-Fe, and $10^{-3} M$ Mg^{2+} and K^+ (Lindsay, 1979).

and Lindsay, 1975; Marshall, 1977). This diagram suggests that 2:1 type clay minerals such as Mg-montmorillonite are very stable minerals at high $H_4SiO_4^0$. Their stability increases with increase in pH. In the intermediate silica range, 1:1 type minerals such as kaolinite are more stable. As $H_4SiO_4^0$ drops below $10^{-5.31}M$, $Al(OH)_3$(gibbsite) becomes the most stable mineral. Thus, during the weathering of primary minerals with concomitant loss of silica, this sequence of secondary weathering products can form. Only at this pH and high activities of K^+ is muscovite more stable than kaolinite while illite is not permanently stable in soils.

The stability diagram in Fig. 7 is based on the selected standard free energies of formation by Sadiq and Lindsay (1979). As new solubility values are reported for aluminosilicate minerals, they, too, can be represented on this diagram and conclusions can be made regarding their stabilities. Measurements on soils can then be made to test the predicted stability relationships shown here.

Phosphate Equilibria

Phosphate is present in soils at approximately 600 ppm (Lindsay, 1979). Since phosphate is an essential nutrient for all living organisms, it has been widely studied. Phosphorus is known to combine with many cations and to form numerous minerals in environments such as soils (Lindsay et al., 1962; Lindsay and Vlek, 1977). How can the solubility relationships of the various phosphate minerals be depicted in a meaningful way?

The stability relationships of several Ca, Fe, and Al phosphates are plotted in Fig. 8 as a function of pH (Lindsay, 1979). At high pH and calcium phosphates decrease in solubility in the order: dicalcium phosphate dihydrate (DCPD) > dicalcium phosphate (DCP) > octacalcium phosphate (OCP) > β-tricalcium phosphate (β-TCP) > hydroxyapatite (HA) > fluorapatite (FA).

In acid soils $FePO_4 \cdot 2H_2O$(strengite) and $AlPO_4 \cdot 2H_2O$(variscite) are more stable. Their solubilities shift, depending on the solubilities of Fe^{3+} and Al^{3+}. For this figure Soil-Fe was used to control Fe^{3+} and kaolinite-quartz to control Al^{3+}. Liming acid soils containing Fe and Al phosphates can be expected to increase phosphate solubility. If the soils are limed to pH > 6.5, Ca phosphates can precipitate and lower phosphate solubility.

Changes in slope of solubility lines at pH 7.2 reflect the fact that $H_2PO_4^-$ is the dominant phosphate ion below this pH, and HPO_4^{2-} is the dominant species above it. The abrupt increase in phosphate solubility above pH 7.8 results from the depression of Ca^{2+} activity by calcite. In acid soils Soil-Ca (Eq. 7) was used to control Ca^{2+} activity. In the presence of calcite, Ca^{2+} activity was controlled according to Eq. 9 with $CO_2(g)$ at $10^{-3.52}$ atm. The reference level for F^- used in Fig. 8 is that in equilibrium with CaF_2(fluorite). These equilibria demonstrate the importance of soil-metal reactions in controlling phosphate solubility. Phosphate minerals

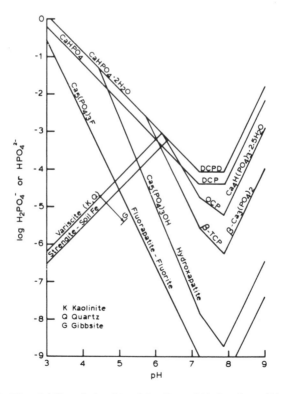

Fig. 8. The solubility relationships of Ca, Fe, and Al phosphates (Lindsay, 1979).

other than those included in Fig. 8 may also be important in soils. These minerals need to be identified, and their solubility relationships need to be measured. Only then can additional phosphate minerals be included on the stability diagram of Fig. 8.

Effect of Redox on Phosphate Solubility

Rice (*Oryza sativa* L.) soils are often flooded to increase the availability of various plant nutrients. For example, reducing a soil has been shown to make $FePO_4 \cdot 2H_2O$(strengite) more soluble (Williams and Patrick, 1971). Can solid phase-solution equilibria be used to demonstrate this expected relationship?

The effect of pe + pH on the stability of $FePO_4 \cdot 2H_2O$(strengite) and its transformation to $Fe_3(PO_4)_2 \cdot 8H_2O$(vivianite) is shown in Fig. 9. The oxidation state of iron in strengite is $+3$ while that in vivianite is $+2$. Changing redox does not affect phosphate solubility at pe + pH > 12.0. Below this redox the transformation of strengite to vivianite is affected by whether Soil-Fe or magnetite controls Fe^{3+} activity. In the presence of vivianite, lowering redox lowers the solubility of phosphate.

If reducing a soil can be expected to depress phosphate solubility, why does submerging acid soils generally increase P availability? In paddy soils where rice is grown, the bulk soil solution can show a very low

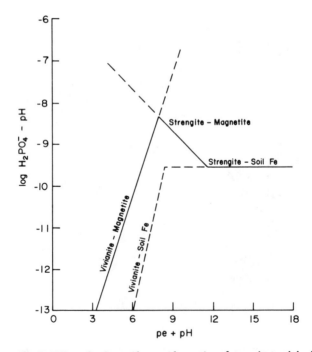

Fig. 9. Effect of redox on the transformation of strengite to vivianite.

phosphate solubility due to the presence of vivianite. However, in the immediate vicinity of rice roots where a more oxidized environment exists, the solubility of vivianite is very much higher so phosphate would be much more available. Seasonal flooding and drying of paddy soils cause $Fe(III)$ phosphates such as strengite to change to $Fe(II)$ phosphates such as vivianite. During the drying period, the soil again oxidizes and vivianite dissolves to precipitate as amorphous Fe phosphate. The amorphous phosphate is generally much more available than crystalline strengite (Huffman, 1962). Furthermore, flooding acid soils usually increases the pH to near 7.0. This can also increase the solubility of strengite (Fig. 8).

Thus an anion like phosphate, which is not directly affected by redox, can be affected indirectly through the effect of redox on the cation with which phosphorus is associated.

Chelate Equilibria

Natural chelates are formed as partial decomposition products of organic matter and as exudates of roots and other living organisms. Although there has been much interest in identifying and characterizing these important constituents in soils, quantitative relationships are still lacking. Fortunately synthetic chelating agents have been used as micronutrient fertilizers and their reactions in soils have been more fully characterized.

The chelate EDTA (ethylenediaminetetraacetic acid) can hold various metal ions in soils over the pH range of 4 to 10 (Fig. 10) (Lindsay, 1979). This graph was generated from the equilibrium reactions of the

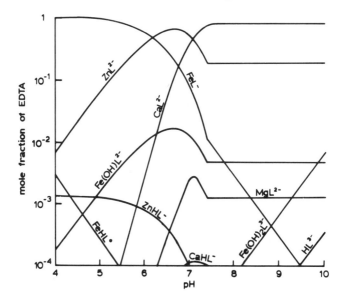

Fig. 10. The mole fraction of EDTA species present in soils when soil Zn^{2+}, Fe^{3+}, Ca^{2+}, Mg^{2+}, and H^+ are the competing metal ions at 0.003 atm of $CO_2(g)$ (Lindsay, 1979).

various soil-metal species (Eq. 4 to 10) and from the formation constants of the various metal-chelate species. The plot represents equilibrium conditions which satisfy all the input equilibria.

At low pH, FeL^- is the major chelated metal ion where L represents the chelating ligand. As pH rises, Fe^{3+} is depressed by the solubility of Soil-Fe (Fig. 5) and Zn^{2+} is able to compete more favorably for the ligand. Although the formation constant for CaL^{2-} is only $10^{-5.83}$ that of ZnL^{2-}, Ca becomes the dominant chelated metal above pH 7 because Zn^{2+} activity decreases to less than $10^{-5.83}$ that of Ca^{2+} (Fig. 2). Above pH of 7.6 calcite controls Ca^{2+} activity and all chelated species having a $^-2$ charge flatten. Equilibrium diagrams such as this are extremely useful to predict metal chelate equilibria in soils.

Similar diagrams can be constructed for other chelates and cations (Norvell, 1972). For EDTA and DTPA (diethylenediaminepentaacetic acid) Lindsay et al. (1967) and Norvell and Lindsay (1969, 1972) showed a close correspondence between predicted and experimentally determined behavior.

Effect of Redox on Chelate Equilibria

Changes in redox affect the activity of many metal ions in soils as demonstrated for Cu and Mn (Fig. 4 and 5). Changes in redox can be expected to change the distributions of metal ions held by chelates.

Redox can affect the stability of metal chelates as demonstrated for EDTA in Fig. 11 (Lindsay, 1979). The plot is shown for pH 7.0 and 0.1

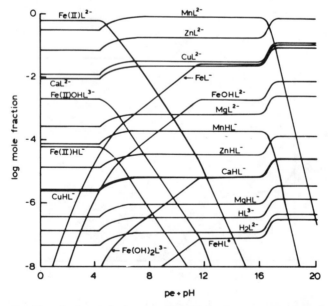

Fig. 11. The effect of redox on the fraction of EDTA that is present as various chelated species (Lindsay, 1979).

atm of $CO_2(g)$ which are common conditions found in submerged soils where aeration is impeded.

As pe + pH drops below 16, Mn^{2+} solubility increases sufficiently (Fig. 5) to displace a significant fraction of other metal ions from the chelate, and MnL^{2-} becomes the predominantly chelated species. Only as pe + pH drops below 4 does Fe^{2+} displace other cations sufficiently that $Fe(II)L^{2-}$ becomes the dominant chelated species (Fig. 11).

Reddy and Patrick (1977) showed experimentally that lowering redox generally displaced Cu and Zn from EDTA and DTPA. Unfortunately, their $CO_2(g)$ measurements were not reported, so it is impossible to compare measured and predicted stabilities. Further studies on the effect of redox on metal chelation have been reported (Sommers and Lindsay, 1979; Abuzkhar and Lindsay, 1981).

CONCLUSIONS

Equilibrium relationships are useful in examining mineral equilibria and solubility relationships in soils. In general, soil scientists have been negligent in making full use of the vast accumulation of thermodynamic data in the chemical literature. This stems in part from the perplexity of trying to select reliable data from the literature and in part the fact that many soil scientists have been trained to believe that equilibrium relationships are of little value in soil systems where so many factors appear as unknowns.

Recently, concerted efforts have been made to screen the literature for reliable solubility data and to make it available to soil scientists. This effort has helped to direct research efforts in soils to fill some of the important missing gaps of information that are so necessary to piece together the overall equilibrium relationships in soils.

Hopefully the examples used in this paper will help to show that the principles of chemistry can apply to soils when solid phases are fully characterized and speciation in the soil solution is resolved. Readers are encouraged to make greater use of equilibrium relationships to predict chemical reactions, mineral stability, and solubility relationships in soils.

LITERATURE CITED

1. Abuzkhar, A. A., and W. L. Lindsay. 1981. Effect of redox on the equilibrium relationships of EDTA in soil. Soil Sci. Soc. Am. J. 45: submitted.

2. Bass Becking, L. G. N., I. R. Kaplan, and D. Moore. 1960. Limits of the natural environment in terms of pH and oxidation-reduction potentials. J. Geol. 68:243–284.

3. Dixon, J. B., and S. B. Weed (ed.). 1977. Minerals in soil environments. Soil Sci. Soc. Am., Madison, Wis.

4. Garrels, R. M., and C. L. Christ. 1965. Solutions, minerals, and equilibria. Harper and Row, New York.

5. Gotoh, S., and W. H. Patrick, Jr. 1972. Transformation of manganese in a waterlogged soil as affected by redox potential and pH. Soil Sci. Soc. Am. Proc. 36:738–742.

6. Huffman, E. O. 1962. Reactions of phosphate in soils: Recent research by TVA. Proceedings 7. The Fert. Soc., London.

7. Hassett, J. J., and J. J. Jurinak. 1971. Effect of Mg^{2+} ion on the solubility of solid car-
    bonates. Soil Sci. Soc. Am. Proc. 35:403–406.

8. Kittrick, J. A. 1971. Montmorillonite equilibria and the weathering environments. Soil
    Sci. Soc. Am. Proc. 36:815–820.

9. Lindsay, A. L. 1972. Zinc in soils and plant nutrition. Adv. Agron. 24:147–186.

10. Lindsay, W. L. 1974. Role of chelation in micronutrient availability. p. 507–524. In
    E. W. Carson (ed.) The plant root and its environment. Univ. Press of Virginia,
    Charlottesville.

11. ————. 1979. Chemical equilibria in soils. Wiley-Interscience, New York.

12. ————, A. W. Frazier, and J. F. Stepheson. 1962. Identification of reaction products
    from phosphate fertilizers in soils. Soil Sci. Soc. Am. Proc. 26:446–452.

13. ————, J. F. Hodgson, and W. A. Norvell. 1967. The physiochemical equilibrium of
    metal chelates in soils and their influence on the availability of micronutrient cations.
    p. 305–316. In G. V. Jacks (ed.) Soil chemistry and fertility. Int. Soc. Soil Sci.

14. ————, and W. A. Norvell. 1969. Equilibrium relationships of Zn^{2+}, Fe^{3+}, Ca^{2+}, and
    H^+ with EDTA and DTPA in soils. Soil Sci. Soc. Am. Proc. 33:62–68.

15. ————, and M. Sadiq. 1981. Use of pe + pH as a redox parameter in soils. Soil Sci.
    Soc. Am. J. 45: submitted.

16. ————, and P. L. G. Vlek. 1977. Phosphate minerals. p. 639–672. In J. B. Dixon and
    S. B. Weed (ed.) Minerals in soil environments. Soil Sci. Soc. Am., Madison, Wis.

17. Marshall, C. E. 1977. The physical chemistry and mineralogy of soils. Vol. 2. Soils in
    place. Wiley-Interscience, New York.

18. McKenzie, R. M. 1977. The manganese oxides and hydroxides. p. 181–193. In J. B.
    Dixon and S. B. Weed (ed.) Minerals in soil environments. Soil Sci. Soc. Am., Madison,
    Wis.

19. Moreno, E. C., W. L. Lindsay, and G. Osborn. 1960. Reactions of dicalcium phosphate
    dihydrate in soils. Soil Sci. 90:58–68.

20. Norvell, W. A. 1972. Equilibria of metal chelates in soils solution. p. 115–138. In J. J.
    Mortvedt, P. M. Giordano, and W. L. Lindsay (ed.) Micronutrients in agriculture.
    Soil Sci. Soc. Am., Madison, Wis.

21. ————, and W. L. Lindsay. 1969. Reactions of EDTA complexes of Fe, Zn, Mn, and
    Cu with soils. Soil Sci. Soc. Am. J. 33:86–91.

22. ————, and ————. 1972. Reactions of DTPA chelates of iron, zinc, copper, and
    manganese with soils. Soil Sci. Soc. Am. Proc. 36:778–783.

23. ————, and ————. 1981. Estimation of iron (III) solubility from EDTA chelate
    equilibria in soils. Soil Sci. Soc. Am. J. 45: submitted.

24. Ponnamperuma, F. N. 1972. The chemistry of submerged soils. Adv. Agron. 24:29–96.

25. ————, T. A. Loy, and E. M. Tianco. 1969. Redox equilibria in flooded soils. II.
    The MnO_2 system. Soil Sci. 108:48–57.

26. Rai, Dhanpat, and W. L. Lindsay. 1975. A thermodynamic model for predicting the
    formation, stability, and weathering of common soil minerals. Soil Sci. Soc. Am. Proc.
    39:991–996.

27. Reddy, C. N., and W. H. Patrick, Jr. 1977. The effect of redox potential on the stability
    of Zn andCu chelates in flooded soils. Soil Sci. Soc. Am. J. 41:429–432.

28. Sadiq, M. 1977. Use of electron titration to study Fe and Mn in soils. Ph.D. thesis. Dep.
    of Agronomy, Colorado State Univ., Fort Collins, Colo. Univ. Microfilm no. 7802397.

29. ————, and W. L. Lindsay. 1979. Selection of standard free energies of formation
    for use in soil chemistry. Colorado Exp. Stn. Tech. Bull. 134.

30. Sillen, L. G., and A. E. Martell. 1971. Stability constants of metal-ion complexes,
    Supplement No. 1, Spec. Publ. No. 25. The Chem. Soc., London.

31. Sommers, L. E., and W. L. Lindsay. 1979. Effect of pH and redox on predicted heavy
    metal-chelate equilibria in soils. Soil Sci. Soc. Am. J. 43:39–47.

32. Stumm, Werner, and J. J. Morgan. 1970. Aquatic chemistry. Wiley-Interscience, New
    York.

33. Truesdell, A. 1969. The advantage of using pE rather than Eh in redox calculations. J. Geol. ed. 17:17–20.

34. Williams, B. G., and W. H. Patrick, Jr. 1971. The effect of Eh and pH on the dissolution of strengite. Nature Phys. Sci. 234(44):16–17.

35. Zordan, T. A., and L. G. Hepler. 1968. Thermochemistry and oxidation potentials of manganese and its compounds. Chem. Rev. 68:737–745.

QUESTIONS AND ANSWERS

Q. On some of your solubility diagrams you showed straight lines labeled Soil-Fe, Soil-Zn, etc. Exactly what is your interpretation of these parameters?

A. These lines were obtained from solubility measurements in soils that are based on the activities of Fe^{3+}, Zn^{2+}, etc. as a function of pH as demonstrated in Fig. 2. Even though we don't know the specific solid phase that controls the solubility of a given metal ion in the soil solution, we can express the activity-pH relationship by an equivalent solubility relationship. In the case of Soil-Zn as shown in Fig. 3, the mineral $ZnFe_2O_4$ (franklinite) is a very distinct possibility of a mineral that controls Zn^{2+} in soils near the observed Soil-Zn level. Further careful experimentation is needed to confirm or refute this hypothesis. In the case of Soil-Fe, our present thinking is that iron precipitates in soils as a ferric hydroxide precipitate that is slightly more soluble (higher free energy) than the recognized crystalline ferric oxides. Slowly with time this more randomly precipitated form of iron oxide may develop greater orderliness as its free energy and consequently its solubility change from amorphous ferric oxide to one of the crystalline forms of ferric oxide.

Q. You use pure minerals in your diagrams, but there are many instances, of course, where substitutions into lattices, say manganese into carbonate lattices, cadmium into carbonate lattices, and others have been implicated in controlling factors too. How do you view that?

A. In developing solubility diagrams such as I have shown, we can make use of any reliable thermodynamic data of solid phases. It so happens that most available reliable data is on pure minerals. Qualitatively we can visualize that when various other ions are substituted into pure minerals, they may not fit perfectly, so the energy of that mixture will likely be a little higher than that of the pure mineral. With time the impurity may tend to escape depending on its activity in the surrounding environment, but these relationships can be worked out providing we have reliable thermodynamic values for the substituted minerals.

Q. In the solubility diagrams that you presented you can see how the inorganic species behave, but how do the organic ligands affect the relationships?

A. The presence of soluble organic ligands in soils can combine with certain inorganic ions to form complexes or chelates that will raise the total soluble inorganic constituents in the soil solution. However, so long as the solid phases controlling the inorganic ions in solution are present, they will still control the inorganic ions as indicated in the diagrams. The organic complexes will increase the mobility of complexed ions and make them more available to plants, etc., but they will not affect the ionic activities in solution. This is a very important principle and simplifies understanding and interpretation of these very useful relationships.

Q. How are you going to be able to deal with the organic complexes?
A. Just as soon as someone will tell us what the organic complexes in soils are and what their equilibrium constants are, we will be able to fit them into our models very easily.

Q. Do you have any evidence that gibbsite is a prerequisite to the formation of variscite in soils?
A. The presence or absence of gibbsite in soils is not the determining factor. There are many soils that do not contain gibbsite. In the weathering sequence where kaolinite-quartz or montmorrillonite-quartz are stable, gibbsite cannot form because Al^{3+} activity is depressed too low. In very acid soils Al^{3+} increases and even though gibbsite cannot form, variscite can. In summary I would say no, gibbsite is not a prerequisite for the formation of variscite, but the presence of gibbsite insures a higher level Al^{3+} and makes its formation more likely.

CHAPTER 11

Factors Affecting the Solubilities of Trace Metals in Soils

S. V. MATTIGOD, GARRISON SPOSITO, AND A. L. PAGE[1]

INTRODUCTION

An improved knowledge of trace metal interactions in soils is essential for evaluating and predicting the behavior of these elements in soil-water systems. Soil systems are a dynamic and complex array of inor-

---

[1] Dep. of Soil and Environmental Sciences, Univ. of California, Riverside, CA 92521.

ganic and organic constituents. At any instant, the trace metal concentrations in the solution phase of such a system are governed by various reactions such as acid-base equilibria, complexation with organic and inorganic ligands, precipitation and dissolution of solids, oxidation-reduction, and ion-exchange-adsorption. The rate at which these reactions occur and the rate of biological uptake together control the concentrations of trace metals in the solution phase. Therefore, reliable predictions of distribution of trace metals within a soil system are dependent on the accuracy with which the effects of various classes of reactions can be predicted. The interrelationships between these reactions in soil systems are indicated schematically in Fig. 1.

TRACE ELEMENT CONCENTRATIONS IN SOIL SOLUTIONS

Typically, elements which are present in the dissolved phase at concentrations less than $10^{-4}M$ are considered to be trace elements. The total dissolved concentrations of some trace elements along with the total concentrations of major elements that are generally encountered in soil solutions are shown graphically in Fig. 2. It is apparent that concentrations of trace elements in soil solutions are generally orders of magnitude less than the concentrations of major elements. The concentration of major elements, therefore, may exert a significant competitive influence on the reactions of trace metal in soils. Both major and trace elements compete in reactions involving complex formation, solid formation, and adsorption onto solid surfaces.

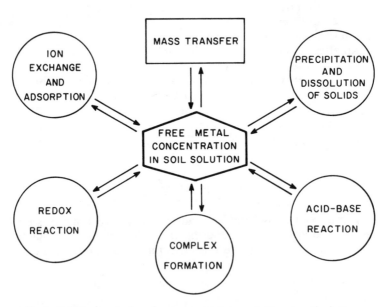

Fig. 1. Principal controls on free trace metal concentrations in soil solutions.

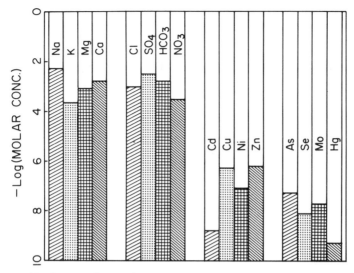

Fig. 2. Typical concentrations of various constituents in soil solutions.

TRACE METAL SPECIATION IN SOIL SOLUTIONS

Trace metals can form soluble complexes with inorganic and organic ligands that are present in soil solutions. These types of reactions can be expressed generally as:

$$aM^z + bL^{t-} = M_aL_b^{az-by} \text{ (aq)} \qquad [1]$$

where

M = Metal ion with charge $z+$
L = Ligand ion with charge $y-$
a and b = Stoichiometric coefficients.

An association constant for reaction 1 is defined by:

$$K_{M_aL_b} = \frac{[M_aL_b]}{[M^{z+}]^a[L^{y-}]^b} \qquad [2]$$

where [ ] denotes an activity.

A number of factors can affect the stability of an aqueous complex relative to its component free ions. Some of these factors are recognized to be the valences of the ionic components, their distance of approach, the degree of covalency in bonding, the number of ligands and cations in the complex, and coordination, polarizability and ligand field effects in case of transition metal complexes. Generally, the stability of complexes increases with cation valence and decreases with ionic radius as predicted by the principles of electrostatics.

A useful concept to explain the observed metal-ligand interactions has been developed by Pearson (1963, 1966). This principle, known as the Hard and Soft Acid and Base (HSAB) theory, groups the metal ions (Lewis acids) and ligands (Lewis bases) into "hard," "soft," and "borderline" categories depending on their polarizability, electronegativity, and oxidation potential. The theory states that the preferences for binding are that hard acids prefer hard bases and soft acids prefer soft bases. Table 1 lists the various metals and ligands grouped according to the HSAB theory and shows that many of the trace metals of the first transition series are in the borderline category. According to HSAB theory, these metals can bind with either hard or soft bases. All of the major cations that are encountered in soil solutions are classified as hard acids and as such would

Table 1. Classification of metals and ligands according to HSAB Theory.[†]

Hard acids	Borderline	Soft acids
Electron configuration of inert gas. Low polarizability: "Hard Spheres"	1–9 outer shell electrons; spherically not symmetrical	Electron number corresponds to NI^0, Pd^0, and Pt^0 (10 to 12 outer shell electrons). Low electronegativity, high polarizability: "Soft Spheres"
(H^+), Li^+, Na^+, K^+, Be^{2+}, Mg^{2+}, Ca^{2+}, Sr^{2+}, Al^{3+}, Sc^{3+}, La^{3+}, Si^{4+}, Ti^{4+}, Ar^{4+}, Th^{4+}, Cr^{3+}, Mn^{3+}, Fe^{3+}, Co^{3+}, UO^{2+}, VO^{2+}	V^{2+}, Cr^{2+}, Mn^{2+}, Fe^{2+}, Co^{2+}, Ni^{2+}, Cu^{2+}, Zn^{2+}, Pb^{2+}, Bi^{3+}	Cu^+, Ag^+, Au^+, Tl^+, Ga^+, Cd^{2+}, Hg^{2+}, Sn^{2+}, Tl^{3+}, Au^{3+}, In^{3+}
as well as species like BF_3, BCl_3, SO_3, RSO_2^+, RPO_2^+, CO_2, RCO^+, R_3C^+	SO_2, NO^+, $B(CH_3)_2$	All metal atoms, bulk metals I_2, Br_2, ICN, I^+, Br^+

Preference for ligand atom	
$N \gg P$	$P \gg N$
$O \gg S$	$S \gg O$
$F \gg Cl$	$I \gg F$

Qualitative generalizations on stability sequence

Cations	Cations
Stability = prop. $\dfrac{\text{Charge}}{\text{Radius}}$	Irving-Williams order $Mn^{2+} < Fe^{2+} < Co^{2+} < Ni^{2+}$ $< Cu^{2+} > Zn^{2+}$
Ligands	Ligands
$F > O > N = Cl > Br > I > S$ $OH^- > RO\text{-} > RCO_2^-$ $CO_3^{2-} \gg NO_3^-$ $PO_4^{3-} \le SO_4^{2-} \gg ClO_4^-$	$S > I > Br > Cl = N > O > F$

Hard bases	Borderline	Soft bases
H_2O, OH^-, F^-, $CH_3CO_2^-$, PO_4^{3-}, SO_4^{2-}, Cl^-, CO_3^{2-}, ClO_4^-, NO_3^-, ROH, RO^-, R_2O, NH_3, RNH_2, N_2H_4	$C_6H_5NH_2$, C_5H_5N, N_3^-, Br^-, NO_2^-, SO_3^{2-}	R_2S, RSH, RS^-, I^-, SCN^-, $S_2O_3^{2-}$, R_3P, R_3As, $(RO)_3P$, CN^-, RNC, CO, C_2H_4, C_6H_6, H^-, R^-

† Based on Pearson (1963), Pearson (1967), and Stumm and Brauner (1973).

prefer to bind with hard bases. Generally in soil systems, complexes involving ligands in the hard base category and metal ions can be expected to be dominant except in cases where reducing conditions and/or significant concentrations of dissolved organic compounds are encountered.

At any instant, the total concentration of each metal ion in the soil solution is the sum of free ion concentration and the concentrations of various metal ion complexes. The mass balance can be expressed as:

$$\text{Tot } M_i = (M_i^{z+}) + \sum_{j=1}^{n} a(M_{ia}L_{jb}^{az-by}) \qquad [3]$$

where

$\text{Tot } M_i$ = Total metal concentration of ith trace metal

(M_i^{z+}) = Free ion concentration of ith trace metal

$(M_{ia}L_{jb}^{az-by})$ = Concentration of complex involving i^{th} trace metal and j^{th} ligand

a and b = Stoichiometric coefficients.

Therefore, a significant factor in controlling total metal concentrations in soil solutions is the concentrations of various ligands present and the stability of the resulting complexes with various metals.

Some probable complexes of a bivalent trace metal ion with some ligands that are hard bases and are likely to be dominant in soil systems are indicated in Table 2. In addition to these complexes, polynuclear species and mixed metal ligand complexes may also exist. The nature of the trace metal complex species in a soil solution influences significantly the extent of interrelated reactions shown in Fig. 1 and, therefore, the mass transfer through the system. For a given total trace metal concentration, the mobility of that metal through a soil system is governed by the free metal concentration, the dominant complex species and the charge on the complexes. These may change continually from point to point within the system depending on the relative importance of the related reactions at each point.

The nature of the dominant complex species in a soil solution may vary significantly among trace metals. For example, the results of a gel filtration experiment by Baham et al. (1978) involving trace metal-fulvic acid solutions indicated that Cd-chloro complexes were dominant among Cd-complexes, whereas, Cu-fulvate complexes were the principal Cu-complex species in the system. Since the total concentrations of the metals were not significantly different in their system, the observed dominance

Table 2. Some probable bivalent metal complexes with inorganic ligands in soil solutions.

OH	Cl	SO$_4$	CO$_3$	PO$_4$
MOH$^+$	MCl$^+$	MHSO$_4^+$	MHCO$_3^+$	MH$_2$PO$_4^+$
M(OH)$_2^0$	MCl$_2^0$	MSO$_4^0$	M(HCO$_3$)$_2^0$	MHPO$_4^0$
M(OH)$_3^-$	MCl$_3^-$	M(SO$_4$)$_2^{2-}$	MCO$_3^0$	MPO$_4^-$
M(OH)$_4^{2-}$	MCl$_4^{2-}$		M(CO$_3$)$_2^{2-}$	

of each complex species can be understood by comparing the stability constant of each species. The reported log K values for CdCl⁺, CdCl₂, and CuCl⁺ species are 1.98, 2.60, and 0.40 respectively (Smith and Martell, 1976). These values indicate that Cd forms more stable complexes with Cl⁻ ions than Cu does and therefore Cd-Cl complex species can be expected to be predominant in this system. The results of this experiment also suggest that Cu-fulvate complexes are significantly more stable than Cd-fulvate complexes.

In another experiment, Doner (1978) studied the mobilities of Cd, Cu, and Ni through soil columns as influenced by Cl⁻ and ClO₄⁻ ions. Leaching experiments with NaClO₄ were included as a reference because significant complexes of ClO₄⁻ with Cd, Cu, and Ni do not exist. The results of this experiment are illustrated in Fig. 3, 4, and 5. The breakthrough curves clearly indicate that the effect of chloride on the mobility of these trace metals was in the order Cd ≫ Cu > Ni. These observed mobilities are in the same order as that of the log K values of association of these metal ions with chloride ion: 1.98 ≫ 0.40 > −0.43.

These experiments clearly indicate that the nature of the metal-ligand complexes significantly affects the trace metal mobilities in a soil system. Therefore, a knowledge of the nature and concentrations of various trace metal species is a prerequisite in understanding the interactions and interpreting the observed mobilities of trace metals in soils.

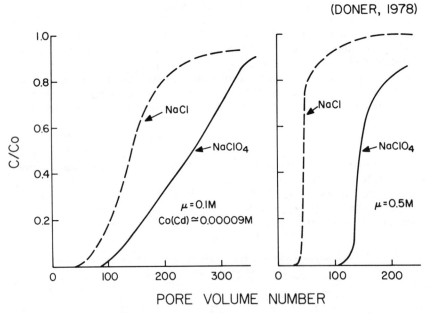

Fig. 3. Breakthrough curves for Cd as affected by Cl⁻ and ClO₄⁻ ions.

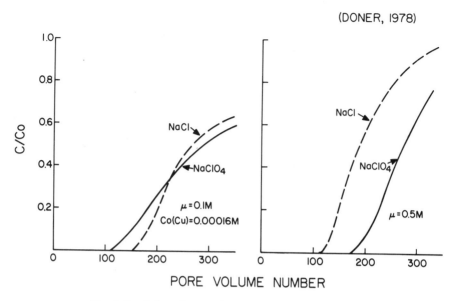

Fig. 4. Breakthrough curves for Cu as affected by Cl⁻ and ClO₄⁻ ions.

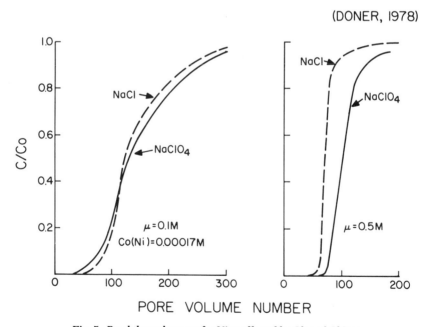

Fig. 5. Breakthrough curves for Ni as affected by Cl⁻ and ClO₄⁻ ions.

DISSOLUTION—PRECIPITATION REACTIONS

Solid—Solution Interactions

In soils, trace metals can occur in a variety of solid phases. At any time in a soil system, with trace metal solids being present, the concentration of trace metals in the solution phase would be influenced by the rate of dissolution of unstable solid phases and the rate of precipitation of stable and metastable solid phases. If we consider a dissolution-precipitation reaction for a solid, M_aL_b:

$$aM^{z+} + bL^{y-} = M_aL_b(s) \qquad [4]$$

M = Metal ion with charge $z+$
L = Ligand ion with charge $y-$
a and b = Stoichiometric coefficients.
The free energy change for this reaction can be expressed as:

$$\Delta G = RT \ln Q/K \qquad [5]$$

where
ΔG = Gibbs energy of reaction ($J\ mol^{-1}$)
R = Gas constant ($J\ K^{-1}\ mol^{-1}$)
T = Absolute temperature (K)
Q = Reaction quotient = $\dfrac{[M_aL_b]}{[M^{z+}]^a[L^{y-}]^b}$
$[\ ]$ = Activity
K = Equilibrium constant = Activity ratio at equilibrium
$\exp(-\Delta G°/RT)$
$\Delta G°$ = Standard Gibbs energy of reaction ($J\ mol^{-1}$).
The direction of reaction (1) is governed by the Q/K ratio, known as the Saturation Index.
If Q = K, ΔG = 0, and the reaction will be at equilibrium.
If Q > K, ΔG > 0, the reaction proceeds to the right, i.e., precipitatation of $M_aL_b(s)$ is favored.
If Q < K, ΔG < 0, the reaction proceeds to the left, i.e., dissolution of $M_aL_b(s)$ is favored.
It should be noted that these thermodynamic relationships indicate the direction of the reaction and not the rate at which it may proceed.

Identification of Solid Phases of Trace Metals in Soils

Positive identification of various solid phases of a trace metal in a soil, along with knowledge of their solubility and their kinetics of dissolution and precipitation, would provide sufficient information to make reliable predictions of trace metal activities in soil solutions. There have been a few successful attempts at identifying the trace metal solids in soils. For example, Norrish (1968) identified in various soils a number of phosphate minerals of the plumbogummite group. Identification of $PbSO_4$ and other Pb compounds that occurred in roadside soils was made by Olson and Skogerboe (1975). Although certain trace metal solids in soils have been identified, the information is very limited and applicable to only a few trace metals. The principal reason for this is that the trace element solids generally constitute a minor part of the total inorganic matrix of soils. For instance, Lindsay and Vlek (1977) have indicated that the average P content of soil is around 0.05 %. Taking into account organic and adsorbed P, the inorganic solid P content would be considerably less than 0.05 %. Thus, a direct identification of trace metal solids in soils presents rather formidable obstacles. In cases where the trace metal solids have been identified directly in soils, the direction of reactions such as (1) for each of the various solids can be determined from a knowledge of the reaction quotient (Q) and the equilibrium constant, K (or $\Delta G°$ for the reaction). Existence in soils, if any, of solid-solution equilibrium can also be recognized.

Equilibrium Solubility Method

DESCRIPTION

Since direct identification of trace metal compounds in soils is a difficult task, an indirect method based on equilibrium solubility has been used extensively by soil chemists. This method consists of comparing the measured ion activity products (IAP) in soil solutions (presumably in equilibrium with minerals) with the equilibrium ion activity products for various solid phases. The minerals whose solubility products are equal to the measured IAP are assumed to be present and hence control the trace metal activities in the soil solution. As an example, Santillan-Medrano and Jurinak (1975) have attempted to relate the observed Pb^{2+} and Cd^{2+} activities in three different soils to the known activity products of some of the Pb- and Cd-bearing minerals. There are, however, a number of limitations to this approach.

LIMITATIONS

To use this method, all minerals that are likely to be present in the soil environment must be known and considered. For example, if the solubility of Ni-carbonate minerals are to be studied, all the known minerals listed in Table 3 have to be included. In addition, it is essential to include other possible solid solutions of Ni-carbonates with other carbonates. Thornber and Nickel (1976) have shown that Ni, Co, Mn, and Cu can form solid solutions of widely-ranging compositions with siderite ($FeCO_3$). This problem was recognized by Jurinak and Santillan-Medrano (1974) in their study of Pb precipitation. They indicated that the likely presence of mixed compounds precluded a more confident interpretation of the data regarding the control of Pb activity in the soil solution.

An additional problem is the lack of equilibrium solubility data (or standard free energy of formation data) for a large number of trace metal

Table 3. Some known Ni-carbonate minerals.†

Name	Formula
Gaspeite	$NiCO_3$
Zaratite	$Ni_3CO_3(OH)_4 \cdot 4\,H_2O$
Otwayite	$(Ni,Mg)_2\,CO_3(OH)_2 \cdot H_2O$
Reevesite	$Ni_6Fe_2CO_3(OH)_{16} \cdot 4\,H_2O$
Takovite	$Ni_6Al_2\,CO_3(OH)_{16} \cdot 4\,H_2O$
Eardleyite	$Ni_{4.92}Mg_{0.10}Ca_{0.02}Fe_{0.13}Al_{2.18}(CO_3)_{2.27}(OH)_{14.42} \cdot 5 \cdot 42\,H_2O$
Carrboydite	$(Ni,Cu)_{6.90}Al_{4.48}(SO_4,CO_3)_{2.78}(OH)_{21.69} \cdot 3 \cdot 67\,H_2O$
Nickelian Calcite	$(Ca,Ni)\,CO_3$
Nickloan hydrozincite	$Ni_{0.37}Zn_{4.63}(CO_3)_2(OH)_6$

† Alwan and Williams (1979).

minerals that are likely to form in soils. This problem can be circumvented to some degree by using empirical methods to estimate the $\Delta G°$ values for those minerals for which experimental data are yet unavailable (Karpov and Kashik, 1968; Kashik, Karpov and Kozlov, 1978; Tardy and Garrels, 1974, 1977; Chen, 1975; Nriagu, 1975, 1976; Tardy and Gortner, 1977; Tardy and Vieillard, 1977; Vieillard, Tardy and Nahon, 1979; Mattigod and Sposito, 1978).

Another limitation to the indirect method is imposed by inaccurate computation of ion activities from the solution data. Direct measurement of free ionic activities by selective-ion electrodes would provide the necessary data to compute IAP. But this is not possible for all ions. The problem arises when free ionic activities are computed inaccurately from the total analytical concentrations of each trace metal and various ligands in soil solutions.

If any of the important ion-pairs and complexes are omitted from computations, the computed free ion activities, and hence the IAP, will be in error. To avoid this problem, the soil solutions should be analyzed not only for the trace metal of interest, but also for the major cations and anions, and for other trace metals which may be present in comparable concentrations. The effects of minor inorganic ligands which may form strong complexes with trace metals as well as soluble organic ligands cannot be ignored. All too frequently, the lack of adequate solution data and omissions in activity calculations lead to erroneous IAP values. These IAP values, coupled with inadequate and/or inaccurate equilibrium solubility data, can lead to erroneous conclusions regarding the trace metal solid phases that may be controlling trace metal solubility.

Lack of knowledge regarding the kinetics of dissolution and precipitation of many trace metal solid phases also precludes a better understanding of the changing chemistry of soil solutions. Soil solution compositions are often controlled by fast-forming metastable solid phases. One well-known example of such a phenomenon is the initial formation of brushite rather than apatite, in soils amended with phosphate fertilizers (Lindsay and Stephenson, 1959). Even though apatite is thermodynamically more stable than brushite, the kinetics of precipitation of the latter phase are much faster than that of apatite. Similarly, the existence of metastable sepiolite, rather than the more stable talc, has been reported in many alkaline soils (Zelazny and Calhoun, 1977). Existence of metastable minerals in soils suggests that in the indirect method, all solid phases that are likely to occur, stable or metastable have to be considered.

Despite these shortcomings, the indirect method may often provide first-approximation information on solid phases that control trace metal activities in soil solutions.

ION EXCHANGE—ADSORPTION REACTIONS

Ion exchange-adsorption reactions play an important role in controlling trace metal concentrations in soil solution. This is particularly true in cases where trace metal concentrations are too low to initiate any precipitation reactions. Also, this type of reaction is generally faster kinetically than many of the precipitation-dissolution reactions and, therefore, can be dominant during periods immediately following an input of dissolved trace element into the soil. In fact, Jenne (1968) proposed that adsorption of trace metals, such as Cd, Ni, Cu, and Zn, onto hydrous oxides of Mn and Fe furnishes the principal control on the concentrations of these metals in soil solutions. From model calculations, Hem (1976) also indicated that Pb levels in solutions may be controlled by ion exchange reactions. However, in many cases, it is impossible to distinguish between adsorption and coprecipitation reactions (Dyck 1968; Hem, 1977; Veith and Sposito, 1977).

There is also some evidence to suggest that adsorption and exchange phenomena for trace metals involve not only the free aquo ions, but also ion-pair and complex species. Species such as $CuCl^+$, $CuOH^+$, Cu-glycine$^+$, $ZnOH^+$, $ZuCl^+$, and Zn-glycine$^+$ can be present as adsorbed or exchangeable species on clay mineral surfaces (Elgabaly and Jenny, 1943; Siegel, 1966; Farrah and Pickering, 1976; Koppelman and Dillard, 1977; Papanicolaou and Nobeli, 1977). Following this line of evidence, a mass balance expression for the total metal adsorbed can be written as:

$$M_T(Ads) = M^{z+} + \sum_{n=1}^{z-1} M(OH)_n^{(z-n)+} + \sum_{i=1}^{x} \sum_{n=1}^{(z-1)/y} ML_{in}^{(z-ny)+} \quad [6]$$

where

$M_T(Ads)$ = Total amount of trace metal absorbed

M^{z+} = Concentration of free aquo ion adsorbed

$M(OH)_n$ = Total concentration of hydrolytic species adsorbed

$ML_{in}^{(z-ny)+}$ [3] Total concentration of all ion-pair and complex species adsorbed

z^+ = Charge on the trace metal cation

n = Number of OH or other ligands

i = i^{th} ligand species

y = Charge on the ith ligand species

It should also be recognized that as adsorption or exchange reactions proceed, the affinity and rate of adsorption of various soluble species will affect the speciation of trace metals and ligands that remain in solution.

There is some direct evidence regarding the adsorption of hydrolytic species of trace metals. Using X-ray photoelectron spectroscopy (XPS),

Koppelman and Dillard (1977) showed that the binding energy of Cu on chlorite was similar to binding energy of Cu in solid $Cu(OH)_2$. Evidence for the adsorption of other trace metal complex species is based principally on observed differences in the quantities of trace metals adsorbed from solution in the presence of different complexing ligands. For example, indirect evidence for the adsorption of $NiNO_3^+$ species on to kaolinite is taken from the work of Mattigod et al. (1979). Figures 6 and 7 show the observed adsorption of Ni by kaolinite from solutions containing $Ca(NO_3)_2$, $CaSO_4$, $NaNO_3$, and Na_2SO_4 as background electrolytes. From the relationship between Ni^{2+} activity in equilibrium solutions and the total Ni adsorbed, the authors concluded that the formation of ion pairs of different stabilities as well as the presence of adsorbing species. Ni^{2+} and $NiNO_3^+$ in NO_3 media, as opposed to only Ni^{2+} in SO_4 media, could explain the observed differences between these adsorption isotherms.

KINETICS OF REACTIONS IN SOIL SYSTEMS

The kinetics of various classes of reactions in soils have an important bearing on trace metal concentrations that appear in soil solutions. It is

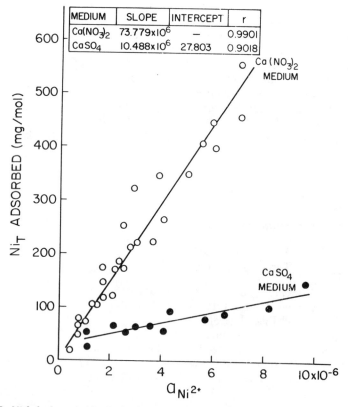

MEDIUM	SLOPE	INTERCEPT	r
$Ca(NO_3)_2$	73.779×10^6	—	0.9901
$CaSO_4$	10.488×10^6	27.803	0.9018

Fig. 6. Nickel adsorption by Ca-kaolinite: Influence of electrolyte and ion pair formation.

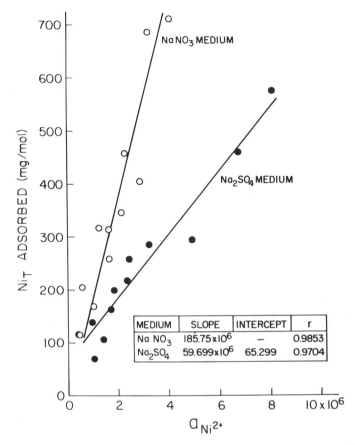

Fig. 7. Nickel adsorption by Na-kaolinite: Influence of electrolyte and ion pair formation.

well-known that the majority of complex formation reactions are very rapid (Wilkins and Eigen, 1965). Therefore, the presumption of equilibrium between free metals, free ligands, and their complexes in soil solutions at all times is highly reasonable.

Many of the ion exchange-adsorption studies involving soil materials have shown either a rapid attainment of equilibrium or a two-step kinetics. Yates (1975) has attributed the first rapid step to adsorption onto highly accessible sites on the adsorbing surface. The second, slower step which may continue for days is characterized by solid state diffusion (Murray, 1975). Therefore, the bulk of the adsorption-exchange reactions can take place within several hours.

In contrast, kinetic data on the dissolution-precipitation reactions of solid phases in soils is extremely sparse. From some of the results published (Hurd et al., 1979; Nancollas et al., 1979; and Plummer et al., 1979) for a few silicates, phosphates, and carbonates, it is evident that kinetic data on the dissolution-precipitation reactions of a large number of solid phases are not yet available. But qualitative data from the geological and geochemical literature suggest that the rates of these types of reactions can vary widely. However, with a knowledge of the existence of various meta-stable and stable solid phases in soil systems, the kinetic factor can be introduced into a predictive scheme such as chemical modeling. Therefore, in the chemical modeling of soil systems the existence of equilibrium with respect to trace metal speciation in solution and ion exchange-adsorption reactions can address adequately the reaction kinetics. The major limitation in modeling is, however, the lack of information about the kinetics of reactions involving solid phases which can be presently treated only qualitatively.

CHEMICAL MODELING OF TRACE METAL EQUILIBRIA IN SOILS

Goals and Problems in Modeling Trace Metal Equilibria

The purpose of chemical modeling of soil systems is to obtain information on the distribution of elements within a soil between solid, aqueous, and gaseous phases at a given point and time. Further, the modeling should be capable of predicting the types and quantities of various solids, the concentrations and distribution of exchangeable and/or adsorbed ions, the metal and ligand speciation in the aqueous phase, and the composition of the gas phase. The changes brought about by mass transfer in the soil system and its impact on changing the partitioning of elements between different phases should be a part of a useful chemical model. To cope successfully with such a task, a model has to account for various classes of reactions (acid-base equilibria, complex formation of metals with organic and inorganic ligands, precipitation and dissolution of solids, redox reactions, and ion exchange-adsorption reactions) that occur in a soil system. The kinetics of reactions should also be part of such a model.

Some of the present limitations on chemical modeling are the lack of stability constant data on numerous soluble complexes, particularly regarding complexes of metals with natural organic compounds, solubility product (or ΔG_f^0) data on various solid phases, a thermodynamic treatment of ion exchange-adsorption reactions, and kinetics data on the dissolution and precipitation of a majority of solid phases. An extensive discussion of the problems, goals, and approaches to chemical modeling has been published by Jenne (1979). Similar questions regarding chemical, physical, and biological models and their interrelationships have been discussed by Morel and Yeasted (1977).

Description of a Model—GEOCHEM

There are a number of chemical equilibrium models capable of simulating the chemical reactions in natural systems. A survey of such models currently in use has been published by Nordstrom et al. (1979). One of these models, GEOCHEM, has been specifically developed for the soil system by Mattigod and Sposito (1979). This computer program was adopted and developed from the REDEQL2 program created originally by Morel et al. (McDuff and Morel, 1974). The program GEOCHEM differs from REDEQL2 in that it i) includes thermodynamic data for hundreds of additional complexes and solids that are particularly relevant to trace metal equilibria in soils; ii) it includes a model for cation exchange on constant charge surfaces; iii) it contains a subroutine for the estimation of single-ion activities in high ionic strength solutions; iv) it has a model to simulate the complexation of trace metals by natural organic compounds.

Both REDEQL2 and GEOCHEM can simulate ion exchange-adsorption reactions based on the James-Healy model (James and Healy, 1972). In addition, both models have the capability to simulate qualitatively, the kinetics of precipitation and dissolution of solid phases. Besides these two models, only one other model (MINEQL) includes this capability.

The development of GEOCHEM has been described by Mattigod and Sposito (1979). Justification for including estimated association constants (Mattigod & Sposito, 1977), and solubility data (Mattigod and Sposito, 1978) for chemical equilibrium computation for multicomponent systems such as soils has also been given.

Applications of GEOCHEM

The use of GEOCHEM to simulate the speciation of trace elements in some sludge-amended soils has been illustrated by Mattigod and Sposito (1979). The results of these computations regarding speciation of Cd, Cu, and Zn were compared with the results of the gel-filtration experiments of Baham et al. (1978). The computed speciation of these metals with respect to soluble organic compounds and Cl⁻ was found to be qualitatively similar to what was observed in the experiments.

Recently, Sposito and Bingham (personal communication) measured the free Cd^{2+} concentrations in the saturation extracts of three soils amended with sewage sludge. These concentrations, measured with a selective-ion electrode for Cd, were compared with the free Cd^{2+} concentrations as computed by GEOCHEM. The computations involved between 200 to 350 soluble complex species were completed in a CPU time of about 70 seconds on an IBM 370/155 computer. The results are indicated in Table 4. The reasonable agreement between measured and calculated values is rather remarkable considering the fact that the computation involved the use of 200 to 350 stability constant values for soluble complexes and that the computer model also simulated the metal complexation by soluble organic compounds in sewage sludge-soil mixtures.

In a series of papers, Batley and Florence (Batley and Florence, 1976; Florence, 1977; Florence and Batley, 1977) have suggested a scheme for classification of heavy metal species in natural waters. This scheme apparently yields the total concentrations of "labile" and "nonlabile" complexes of organic and inorganic ligands with trace metal ions. In this scheme, the use of ion exchange resins and treatment with

Table 4. Comparison between calculated and measured concentrations of Cd^{2+} in saturation extracts of sludge-amended soils. [†]

Saturation extract no.	Hanford		Redding		San Miguel	
	Calcd.	Obsd.	Calcd.	Obsd.	Calcd.	Obsd.
	mmol m⁻³					
1	0.02	<0.1	0.02	<0.1	0.2	<0.1
2	0.9	1.2	0.32	0.68	0.24	0.54
3	1.2	1.0	0.66	1.1	0.42	0.68
4	3.6	3.2	1.6	1.8	0.9	1.2
5	10.4	8.0	3.7	3.7	3.1	3.5
6	19.7	16.0	10.2	11.0	6.0	7.4
7	58.1	35.0	45.5	44.0	21.0	20.0

[†] Sposito and Bingham (personal communication).

buffers can alter the speciation present in natural water samples and thus the results obtained cannot be compared with the results of computer modeling. The method, however, can provide some general insight into the probable existence of some groups of complex species.

Computer modeling of trace metal equilibria, therefore, can provide a useful framework to understand the nature of the multitude of reactions that take place in dynamic multicomponent systems such as soils. The reliability of the model predictions can be constantly improved by updating thermodynamic data in consonance with directly measured experimental parameters. Further improvement and refinement in modeling studies can lead to a better understanding of chemical processes that occur in soil systems.

SUMMARY

Trace metal equilibria in soils can be viewed in terms of reactions within aqueous solution, solid, and gaseous phases. The interaction between these phases and the rate at which these reactions occur and the mass transfer reactions exert control over the trace metal concentrations at any point and time. Chemical models which can account adequately for various classes of reactions can fairly well predict solution phase equilibrium. However, the predictions of chemical models can be only qualitative at best when handling reactions involving ion exchange-adsorption and solid phases. Despite these problems, chemical modeling is a useful framework for a better understanding of trace metal interactions in multicomponent dynamic systems such as soils.

LITERATURE CITED

1. Alwan, A. K., and P. A. Williams. 1979. Nickeloan hydrozincite: A new variety. Min. Mag. 43:397–398.

2. Baham, J., N. B. Ball, and G. Sposito. 1978. Gel filtration studies of trace metal-fulvic acid solutions extracted from sewage sludge. J. Environ. Qual. 7:181–188.

3. Batley, G. E., and T. M. Florence. 1976. A novel scheme for the classification of heavy metal species in natural waters. Anal. Lett. 9:379–388.

4. Chen, C. H. 1975. A method for estimation of standard free energies of formation of silicate minerals at 298.15 K. Am. J. Sci. 275:801–817.

5. Doner, H. E. 1978. Chloride as a factor in mobilities of Ni(II), Cu(II), and Cd(II) in soil. Soil Sci. Soc. Am. J. 42:882–885.

6. Dyck, W. 1968. Adsorption and coprecipitation of silver on hydrous ferric oxide. Can. J. Chem. 46:1441.

7. Elgabaly, M. M., and H. Jenny. 1943. Cation and anion interchange with zinc montmorillonite clays. J. Phys. Chem. 47:399–408.

8. Farrah, H., and W. F. Pickering. 1976. The sorption of copper species by clays. I. Kaolinite. Aust. J. Chem. 29:1167–1176.

9. Florence, T. M. 1977. Trace metal species in fresh waters. Water Res. 11:681–687.

10. ————, and G. E. Batley. 1977. Determination of the chemical forms of trace metals in natural waters, with special reference to copper, lead, cadmium, and zinc. Talanta Mini-Review. Talanta. 24:151–158.

11. Hem, J. D. 1976. Geochemical controls on lead concentrations in stream water and sediments. Geochim. Cosmochim. Acta. 40:599–609.

12. ————. 1977. Reactions of metal ions at surfaces of hydrous iron oxide. Geochim. Cosmochim. Acta. 41:527–538.

13. Hurd, D. C., C. Fraley, and J. K. Fugate. 1979. Dissolution kinetics of silicate rocks-application to solute modeling. p. 447–474. In E. A. Jenne (ed.) Chemical modeling in aqueous systems. ACS Sym. Ser. 93. Am. Chem. Soc.

14. James, R. O., and T. W. Healy. 1972. Adsorption of hydrolysable metal ions at the oxide-water interface. III. A thermodynamic model of adsorption. J. Coll. Int. Sci. 40: 65–81.

15. Jenne, E. A. 1968. Controls on Mn, Fe, Co, Ni, Cu, and Zn concentrations in soils and water: the significant role of hydrous Mn and Fe oxides. p. 337–387. In R. F. Gould (ed.) Trace inorganics in water. Adv. Chem. Sev. 73. Am. Chem. Soc., Washington, D.C.

16. ————. 1979. Chemical modeling-goals, problems, approaches, and priorities. p. 1–21. In E. A. Jenne (ed.) Chemical modeling in aqueous systems. SCS Sym. Ser. 93. Am. Chem. Soc.

17. Jurinak, J. J., and J. Santillan-Medrano. 1974. The chemistry and transport of lead and cadmium in soils. Res. Rep. 18. Utah Agric. Exp. Stn., Logan, Utah.

18. Karpov, I. K., and S. A. Kashik. 1968. Computer calculation of standard isobaric-isothermal potentials of silicates by multiple regression from a crystallochemical classification. Geochem. Int. 5:706–713.

19. Kashik, S. A., I. K. Karpov, and G. V. Kozlova. 1978. An empirical method of calculating free energies for layer silicates. Geochem. Int. 15:70–74.

20. Koppelman, M. H., and J. G. Dillard. 1977. A study of the adsorption of Ni(II) and Cu(II) by clay minerals. Clays Clay Min. 25:457–462.

21. Lindsay, W. L., and H. F. Stephenson. 1959. Nature of the reactions of monocalcium phosphate monohydrate in soils: I. The solution that reacts with the soil. Soil Sci. Soc. Am. Proc. 23:12–18.

22. ————, and P. L. G. Vlek. 1977. Phosphate minerals. Chap. 17. p. 639–672. In Minerals in soil environment. Soil Sci. Soc. Am., Madison, Wis.

23. Mattigod, S. V., and G. Sposito. 1977. Estimated association constants for some complexes of trace metals with inorganic ligands. Soil Sci. Soc. Am. J. 41:1092–1097.

24. ————, and ————. 1978. Improved method for estimating the standard free energies of formation of smectites. Geochim. Cosmochim. Acta. 42:1753–1762.

25. ————, and ————. 1979. Chemical modeling of trace metal equilibria in contaminated soil solutions using the computer program GEOCHEM. p. 837–856. In E. A. Jenne (ed.) Chemical modeling in aqueous systems. ACS Sym. Ser. 93. Am. Chem. Soc.

26. ————, A. S. Gibali, and A. L. Page. 1979. Effect of ionic strength and ion pair formation on the adsorption of nickel by kaolinite. Clays Clay Min. 27:411–416.

27. McDuff, R. E., and F. M. M. Morel. 1974. Description and use of the chemical equilibrium program REDEQL2. Tech. Rep. EQ-73-02. W. M. Keck Lab. Cal Tech.

28. Morel, F. M., and J. G. Yeasted. 1977. On the interfacing of chemical, physical, and biological water quality models. p. 253–268. *In* Fate of pollutants in the air and water environments. V8. John Wiley and Sons.

29. Murray, J. W. 1975. The interaction of metal ions at the manganese dioxide-solution interface. Geochim. Cosmochim. Acta. 39:505–519.

30. Nancollas, G. H., Z. Amjad, and P. Koutsoukas. 1979. Calcium phosphates—speciation, solubility, and kinetic considerations. p. 475–498. *In* E. A. Jenne (ed.) Chemical modeling in aqueous systems. ACS Sym. Ser. 93 Am. Chem. Soc.

31. Nordstrom, D. K. et al. 1979. Comparison of computerized models for equilibrium calculations in aqueous systems. p. 857–894. *In* E. A. Jenne (ed.) Chemical modeling in aqueous systems. ACS Sym. Ser. 93. Am. Chem. Soc.

32. Norrish, K. 1968. Some phosphate minerals of soils. Trans. 9th Int. Congr. Soil Sci. American Elsevier Co., N.Y. 2:713–723.

33. Nriagu, J. O. 1975. Thermochemical approximations for clay minerals. Am. Mineral. 60:834–839.

34. ————. 1976. Phosphate-clay mineral relations in soils and sediments. Can. J. Earth Sci. 13:717–736.

35. Olson, K. W., and R. K. Skogerboe. 1975. Identification of soil lead compounds from automotive sources. Env. Sci. Tech. 9:227–230.

36. Papanicolaou, E. P., and C. Nobeli. 1977. A contribution to the study of $ZnCl^+$ adsorption by soils. Z. Pflanzenernaehr. Bodenkd. 140:543–548.

37. Pearson, R. G. 1963. Acids and bases. Science 151:1721–1727.

38. ————. 1966. Hard and soft acids and bases. J. Am. Chem. Soc. 85:3533–3539.

39. Plummer, L. N., T. M. L. Wigley, and D. L. Parkhurst. 1979. Critical review of the kinetics of calcite dissolution and precipitation. p. 537–576. *In* E. A. Jenne (ed.) Chemical modeling in aqueous systems. ACS Sym. Ser. 93. Am. Chem. Soc.

40. Santillan-Medrano, J., and J. J. Jurinak. 1975. The chemistry of lead and cadmium in soil: Solid phase formation. Soil Sci. Soc. Am. Proc. 39:851–855.

41. Siegel, A. 1966. Equilibrium binding studies of Zn-glycine complexes to ion exchange resins and clays. Geochim. Cosmochim. Acta. 30:757–768.

42. Smith, R. M., and A. E. Martell. 1976. Critical stability constants. V. 4. Inorganic complexes. Plenum Press, New York.

43. Stumm, W., and P. A. Branauer. 1975. Chemical speciation. Chap. 3, p. 173–239. *In* J. P. Riley and G. Skirrow (ed.) Chemical oceanography 2nd ed. Academic Press. London.

44. Tardy, Y., and R. M. Garrels. 1974. A method of estimating the Gibbs energies of formation of layer silicates. Geochim. Cosmochim. Acta. 38:1,101–1,116.

45. ————, and ————. 1977. Prediction of Gibbs energies of formation of compounds from the elements—II. Monovalent and divalent metal silicates. Geochim. Cosmochim. Acta. 41:87–92.

46. ————, and L. Gortner. 1977. Relationships among Gibbs energies of formation of sulfates, nitrates, carbonates, oxides, and aqueous ions. Contrib. Mineral. Petrol. 63:89–102.

47. ————, and P. Vieillard. 1977. Relationships among Gibbs free energies and euthalpies of formation of phosphates, oxides, and aqueous ions. Contrib. Mineral. Petrol. 63:75–88.

48. Thornber, M. R., and E. H. Nickel. 1976. Supergene alteration of sulfides. II. The composition of associated carbonates. Chem. Geol. 17:45–72.

49. Veith, J. A., and G. Sposito. 1977. On the use of the Langmuir equation in the interpretation of "adsorption" phenomenon. Soil Sci. Soc. Am. J. 41:697–702.

50. Vieillard, P., Y. Tardy, and D. Nahon. 1979. Stability fields of clays and aluminum phosphates: paragenesis in lateritic weathering of agrillaceous phosphatic sediments. Am. Mineral. 64:626–634.

51. Wilkins, R. G., and M. Eigen. 1965. The kinetics and mechanism of formation of metal complexes. p. 55–66. *In* R. F. Gould (ed.) Mechanisms of inorganic reactions. Adv. Chem. Ser. 49. Am. Chem. Soc.

52. Yates, D. E. 1975. The structure of oxide/aqueous electrolyte interface. Ph.D. thesis. Univ. of Melbourne, Australia.

53. Zelazny, L. W., and F. G. Calhoun. 1977. Palygorskite (attapulgite), sepiolite, talc, pyrophyllite, and zeolites. Chap. 13. p. 435–470. *In* Minerals in soil environments. Soil Sci. Soc. Am., Madison, Wis.

CHAPTER 12

External Phosphorus Requirements of Crops[1]

R. L. FOX[2]

[1] Hawaii Agric. Exp. Stn. J. series no. 2493.

[2] Professor of Soil Science, Dep. of Agronomy and Soil Sci., Univ. of Hawaii, Honolulu 96822.

INTRODUCTION

The nutrient requirement of a crop can be expressed in several ways. For example, the term "internal nutrient requirement" may mean the minimum uptake of nutrient (a quanity factor in plant nutrition) that is associated with a specified yield. The internal requirement can also be defined as the concentration of a nutrient in the plant (an intensity factor in plant nutrition) that is associated with near maximum yield, usually named the "critical concentration." Crops have external nutrient requirements too. These may be defined as the quantity of nutrient (or some proportional part of that quantity) that constitutes a minimum pool for adequate crop nutrition. But the "external requirement" can also refer to the intensity of nutrition: the concentration of nutrient in the soil solution which is associated with adequate nutrition. The quantity of nutrient frequently takes on greater significance if it is considered in relation to the capacity of the soil to hold nutrients. Thus, both the internal and the external nutrient requirement can be expressed in terms of quantity, intensity, and capacity.

The concept of a nutrient pool or reservoir of soil P from which crops draw has been in vogue for many years. Hilgard, Truog, Morgan, Bray and Kurtz, Mehlich, Olsen, Fried and Dean, is only a partial listing, for the U.S. alone, of familiar names associated with methods that were designed to evaluate such a pool. The "soil solution" concept has been around for many years also. But, although there has been willingness to affirm that the soil solution is important in plant nutrition, very few investigations have proceeded as if this were so. The emphasis has consistently been on capacity and quantity factors and not on intensity factors.

During the past decade, considerable effort has been expended to determine the external P requirements of cultivated crops. The purpose of this paper is to explain the importance of the external P requirement, approaches used to determine it, and how external P requirements can be combined with phosphate sorption curves of soils to determine fertilizer P requirements.

THE SOIL SOLUTION AND P NUTRITION

Intensity, Quantity, Capacity, and Rate Factors

The soil solution provides an important, immediate P source for plants growing in soils. Phosphate in soil solutions, or in dilute salt solutions equilibrated with soils, is closely related to the P nutrition of plants. In 1923, Burd and Martin (1924) introduced a water-displacement method to obtain soil solutions. They concluded that equilibrium between P in the liquid and solid phases is established so rapidly that changes in the moisture content of soils do not appreciably change the P concentration in solution (a conclusion which has required some modification). They also noted that P concentration in solution decreased as a result of cropping. (We now understand that P sorption by soils is concentration-dependent.) Translated into practical soil fertility, solution P concentration is decreased by crop uptake.

The concentration of P in solution is an estimate of the intensity of P nutrition. Since P in soil solutions is extremely dilute, it must be continuously renewed; if not, concentrations will decrease rapidly as soil P is used by plants. A high flux of P to roots is possible, even when P concentrations are low, if solutions bathing root surfaces are quickly and continuously renewed with P. In general, this requires a short diffusion path and a large cross-sectional area through which ions may diffuse; both are a function of soil water contents. For excised maize (*Zea mays* L.) roots, release of solid phase P to the soil solution does not appear to limit the P absorption mechanism (Olsen and Watanabe, 1963).

Usually the source of P that resupplies soil solutions is the labile fraction—the solid phase P that equilibrates with the solution phase—which is a quantity factor in plant nutrition. Attempts have been made to determine the relative importance of quantity and intensity factors. For example, when soils were intensively cropped with ryegrass, the quantity of P became increasingly important as time progressed (Holford and Mattingly, 1976). However, extrapolation of these data suggests that intensity is the dominant factor influencing the P nutrition of ryegrass (*Lolium multiflorum* Lam.) in unstressed soils (Fig. 1). I believe that the unstressed condition is more indicative of field conditions than the condition which results from prolonged cropping in the narrow confines of a small pot. In pot studies maximum short-term plant yields are obtained when soils with high sorption maxima, as measured by the Langmuir isotherm, are about 25% P-saturated (Woodruff and Kamprath, 1965). Except for quartz sands and some organic soils devoid of mineral matter, soils which have adequate P in solution usually do not lack for P quantity per se—at least not in the short term. But if the quantity of P in the labile pool is small in relation to the capacity of soils to sorb P the intensity of P nutrition may not be adequate for maximum crop performance (Cole and Olsen, 1959).

The P flux to plant roots, a rate factor in P nutrition, is the product of numerous secondary rate factors. The dissolution of P compounds, desorption of sorbed P, diffusion of P in soil solutions, and removal of P from solution by roots are examples of important secondary factors which are difficult to evaluate in soils.

Minimum P Concentration for Maximum Crop Growth

Numerous attempts have been made during the past 60 years to determine minimum soil solution P concentrations that will support maximum plant growth. Pierre and Parker (1927) investigated the P content of

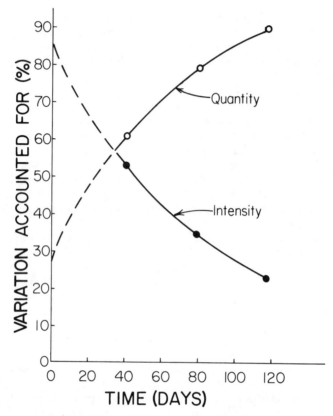

Fig. 1. Ryegrass growth in relation to initial P quantity and P intensity in the soil. Based on data from Holford and Mattingly (1976).

displaced solution from 20 soils. The average concentration of inorganic P was about 0.025 ppm. In solution cultures, maximum growth of maize occurred at a P concentration of about 0.05 to 0.07 ppm P (Parker, 1927). Tidmore (1930) reported that corn, sorghum [*S. bicolor* (L.) Moench.] and tomatoes (*Lycopersicon esculentum* Mill.) will not make satisfactory growth with less than about 0.03 ppm P in solution. Field corn yields of 2.6 and 3.5 tons/ha were associated with soil solution concentrations of 0.01 and 0.013 ppm P. Bingham (1949) investigated the growth of Romaine lettuce (*Lactuca sativa* var. longifolia) in pots in relation to the concentrations of P in water extracts of many California soils. On the average, little evidence was obtained for P response when 1:10 water extracts contained more than 0.4 ppm P.

Plants grown in flowing solution-cultures varied greatly in the concentrations of P they required (Asher and Loneragan, 1967). Some species required only 0.03 ppm P in solution while others required 25 times more.

PHOSPHATE SORPTION CURVES AND THEIR UTILITY

P Sorbed at Standard Concentration

The problems associated with P infertility of soils have been approached in such an arbitrary manner that it is almost impossible to transfer information from one area to another. Progress was made toward that goal when Beckwith (1965) suggested that P adsorption curves could be used to estimate fertilizer requirements. There has been some confusion about Beckwith's proposal, and also about subsequent work based on it. Beckwith suggested a standard concentration of 0.2 ppm P in solution to compare P sorption by soils because 0.2 ppm is an adequate concentration of P in solution for most species. Beckwith did not propose 0.2 ppm as a universal external P requirement for crops—most crops require less.

A Standard Method for Preparing P Sorption Curves

The procedure used by Beckwith to determine P sorption curves was involved, and more artificial than necessary. The method suggested here more nearly corresponds to real field situations. Equilibration time and shaking procedures are such that the fast reaction between P and soil has subsided before P concentration in the equilibrated solution is determined. Separate soil samples are equilibrated at 25 C in 0.01 M $CaCl_2$ containing graduated quantities of $Ca(H_2PO_4)_2 \cdot H_2O$. A 1:10 soil to solution ratio is convenient. Systems have been equilibrated for 6 days with shaking for two 30-min periods each day (Fox and Kamprath, 1970), but this procedure may not be appropriate for all soils. Figure 2A presents the basis for selecting a 6-day equilibration procedure. If soils are continuously agitated, the fast reaction may subside more quickly—even in 1 or 2 days. Care should be exercised that fresh mineral surfaces, upon which P may be adsorbed, are not generated through abrasion during the process of shaking. Phosphorus sorption-desorption curves usually show a hysteresis effect as in Fig. 2B. The effect is exaggerated when wide ratios of soil to equilibrating solution are used.

When P sorption values are plotted against P concentration in equilibrated solutions, curves of the type presented in Fig. 3 are obtained. These curves illustrate the general relationship between soil mineralogy and phosphate retention. The Andepts represent hydrated Fe and Al oxides, the Oxisols represent kaolin clays with a high percentage of Fe and Al minerals, the Ultisols are mostly kaolin type clays and quartz, and the Chernozems are mostly 2:1 silicate clay and quartz. In some cases (examples not presented), an appreciable quantity of P appears in solution when none has been added. This is assumed to have been desorbed and is plotted on the "P sorbed" axis as a negative value. These sorption curves can be used to predict quantities of fertilizer required to establish desired

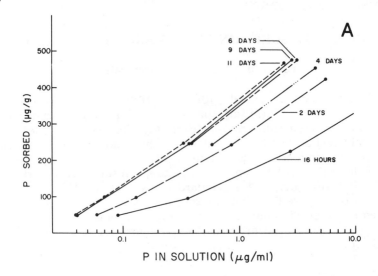

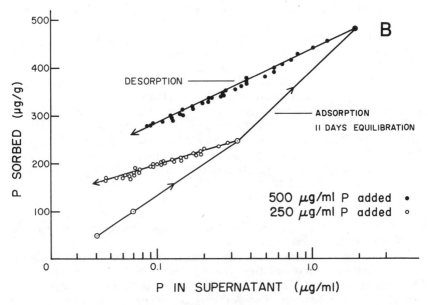

Fig. 2. Phosphate sorption (A) and desorption (B) by Georgeville soil suspended in a sodium acetate solution (Fox and Kamprath, 1970).

concentrations of P in soil solutions for a given volume of soil of a known bulk density. Numerous examples of P sorption curves and the utility of the Standard P Requirement have been published (Fox and Kamprath, 1970; Fox, 1974; Juo and Fox, 1977; Fox and Searle, 1978; Kang and Juo, 1979; Mokurenye, 1979; and Vander Zaag et al., 1979).

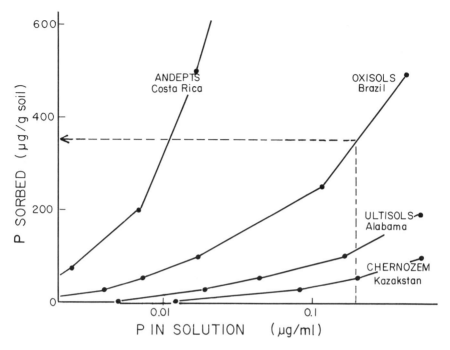

Fig. 3. Composite phosphate sorption curves for selected soils.

THE EXTERNAL P REQUIREMENT

Methods of Approach

PRELIMINARY STUDIES

Short-term pot studies were performed to see if pearl millet (*Pennisetum thyphoides*) had a similar solution P requirement on different soils. Soils were selected which varied in texture (4 to 40% clay) and in amount of fertilizer which had previously been applied (none to 700 kg P/ha). Phosphate sorption curves were developed by the method described above for each soil-treatment combination. These curves were then used to predict the quantities of P needed to adjust the concentration of P in solution to six different levels for each soil. The required P was mixed with the soil, equilibrated, and thoroughly mixed again. Actual concentrations of P in saturation extracts corresponded reasonably well to intended values during early plant growth. Plants were harvested after only 4 weeks growth to minimize the barrier effects of the pot walls. Yield of millet approached 95% of maximum attainable when the intended concentration of P in solution was about 0.2 ppm, irrespective of soil series, soil texture or past P fertilizer treatments (Fox and Kamprath, 1970). The predicted concentration of P in solution at which 95% of

$\sim 7 \mu M$

attainable yield was realized (0.2 ppm in this case) determined by the standard set of laboratory conditions described was named the external P requirement.

The experiment demonstrated that P in solution was the feature of prime importance; texture, mineralogy, residual fertilizer were important in that they influenced the quantity of P required to establish a given solution P concentration.

FIELD EXPERIMENTATION

Some preliminary considerations—A curve of P absorption by roots as a function of P concentration must be sigmoid, with a threshold concentration, below which there is no net P uptake, and a saturation concentration, above which P uptake does not increase. There is reason to believe that plant roots become saturated with respect to P uptake in the approximate range of 0.3 to 1 ppm P in solution (Edwards and Barber, 1976). This value is surprisingly low, suggesting that initial P solubility in most fertilizer bands exceeds the P saturation value for roots. If this is so, P placed in unmixed bands is inefficiently used under most conditions, certainly so in a short term. The results of Yost et al. (1979) have been used to support the validity of this conclusion in a tropical environment (Fox, 1978).

Having made a case for mixing P with at least some soil, the extreme case, mixing a small quantity of P in a large volume of extremely P-deficient and P-retentive soil, should be avoided (Leon and Fenster, 1979). For if the soil's initial P concentration is below the threshold concentration, increasing that concentration with fertilizer to the threshold level theoretically will not increase P uptake. And, although the first increment of P fertilizer applied after the threshold has been reached may be efficiently used, the overall efficiency may still be low (Fox, 1978).

Phosphate applied as unmixed bands (zero mixing) has often been compared with P completely incorporated in the soil. Good examples of field experimentation comparing P incorporated in intermediate volumes are rare. Work in the tropics suggests that if P is adequately supplied, the volume of soil into which it is incorporated is of minor importance (Fox and Kang, 1978; Fox, 1978).

Results of permanent field plots—Field plots were needed in which P could be maintained (within acceptable limits) at set concentrations in the soil solution. Plots were established at six sites having a wide range of edaphic and climatic conditions. Fertilizers were added according to requirements indicated by P sorption curves so that P in equilibrated solution (0.01 M $CaCl_2$) increased in 6 to 10 increments from the original unfertilized condition to 1.6 ppm. A succession of crops was grown. The P concentration of soil solutions decreased as a function of time after fertilization. To counteract that trend, soil solution levels were readjusted to the assigned levels in solution before each new planting. After several rounds of heavy fertilization, P in displaced water solutions

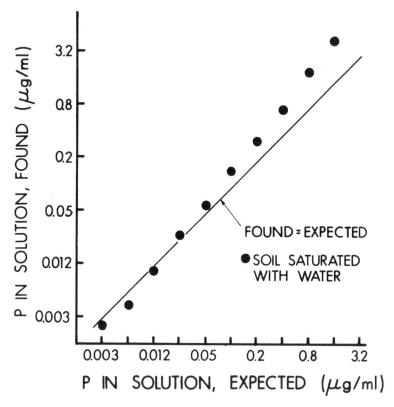

Fig. 4. Relationship between P concentrations found and those expected in solutions displaced from soil columns prepared from samples of Eutrustox soil which had been variously fertilized for 9 years.

was higher than expected (Fig. 4) probably because the equilibration procedure used for preparing the P sorption curves disrupts aggregates, exposing surfaces which do not usually react in the field. Some of the yield response curves from these plots have been published (Fox et al., 1974; Nishimoto et al., 1975, 1977). The examples presented in Fig. 5A clearly demonstrate that there are species differences in the external P requirements of crops. In this case, groundnuts (*Arachis hypogaea*) were relatively insensitive to P in the soil solution while soybean (*Glycine max*), and a forage legume, *Desmodium aparines*, required more P in solution. Table 1 presents the external P requirements for several crop species.

The external P requirement is not a single-valued constant that holds for all conditions but the results do suggest that the concentration of P in dilute salt solutions which have been equilibrated with the soil under standard conditions is a useful indicator of the P nutrition of crops and that the external P requirement might be widely applied in conjunction with P sorption curves to estimate P fertilizer requirements.

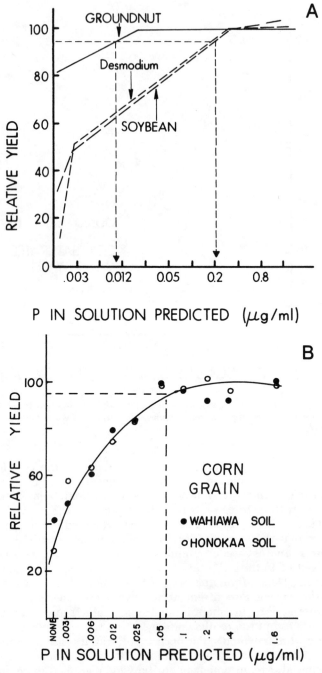

Fig. 5. A. External P requirements of groundnuts and soybeans grown on a Gibbsihumox and Desmodium grown on a Hydrandept (Fox, 1978). B. Corn grain yields in relation to adjusted concentrations of P in two soils with different mineralogical systems (Fox et al., 1974).

Table 1. Concentration of P in solution associated with 95 % of maximum yield of selected crops grown in Hawaii.

Crop	Soil series	Approximate P in soil solution for yield indicated	
		75 % of max.	95 % of max.
		ppm	
Cassava (Manihot esculenta)	Halii	0.003	0.005
Peanut (Arachis hypogaea)	Halii	0.003	0.01
Cabbage (Brassica oleracea)	Kula	0.012	0.04
Corn (Zea mays)	Halii	0.005	0.05
	Wahiawa	0.01	0.06
Sorghum bicolor (grain)	Honokaa	0.015	0.06
Soybean (Glycine max)	Wahiawa	0.025	0.20
	Halii	0.025	0.20
Tomato (Lycopersicon esculentum)	Kula	0.05	0.20
Head lettuce (Lactuca sativa)	Kula	0.10	0.30

Factors Which Influence the External P Requirement

EDAPHIC FACTORS

Field experimentation confirmed the evidence of pot experiments (Fox and Kamprath, 1970) that such edaphic factors as texture and mineralogy had little influence on the external P requirement of crops. The external P requirement for corn was almost identical for corn grown on two dissimilar soils, one an Oxisol and the other a Hydrandept (Fig. 5B). The Wahiawa soil is a Eutrustox, mostly kaolin clay-iron oxide mineralogy, while the Honokaa soil is a Hydrandept with highly hydrated (thixotropic) oxides of iron and aluminum (Fox et al., 1974). The external P requirement at a third location (a Gibbsihumox) was slightly lower (0.05 ppm), while corn growing on a Haplustoll with original P in solution in excess of 0.05 ppm did not respond to P.

The work of Tidmore (1930), referred to earlier, in which corn yields of 2.6 and 3.5 tons/ha were associated with soil solution concentration of 0.01 and 0.013 ppm, agrees remarkably well with expected yields from those concentrations as indicated in Fig. 5B. These experiments suggest that the external P requirement for field production of a particular crop is reasonably constant for a range of soil conditions.

CULTIVAR DIFFERENCES

The search among commercial cultivars for within-species tolerance to low P availability has not been very encouraging (Fox, 1979). Limited experience suggests that some corn varieties from tropical America tolerate P deficiency better than those developed for high-input agriculture of the temperate zone. Other comparisons of within-species variability indicates remarkable uniformity; see, for example, the data for lettuce (Lactuca sativa L.) in Nishimoto et al. (1977).

EFFECT OF WEATHER

Deficient levels of certain other growth factors may put such a low ceiling on yield that the external P requirements of crops are decreased below the normal level (Fox, 1979). For example, in Nigeria, when corn was grown during the middle of the rainy season (low light intensity) the yield potential was 3.8 tons grain/ha and the external P requirement was about 0.01 ppm. At the beginning of the rainy season (4.5 tons/ha yield potential) the P requirement was 0.025 ppm (Fox and Kang, 1978). In Hawaii, a yield potential of 7 tons was associated with 0.05 ppm P. And in northern Idaho where the days are long and bright, sweet corn required about 0.13 ppm P (Jones and Benson, 1975).

Plants which are under low temperature stress apparently have high external P requirements. For example, the external P requirement of winter-grown lettuce in Arizona (soil temperatures decreasing to 13 C) is higher (personal communication, B. R. Gardner, Univ. of Arizona, Yuma) than in Hawaii where the winter soil temperature is about 20 C. Low soil temperature depresses the concentration of P in solution in equilibrium with soils. If low soil temperature coincides with high P demand by crops, plant growth could be adversely affected on soils with marginal P levels (Gardner and Jones, 1973).

MYCORRHIZAL EFFECTS

Mycorrhiza-forming fungi may affect the P nutrition of crops. If the mycorrhiza-forming fungi are eliminated from a P-deficient soil by fumigation, crops which would otherwise make acceptable growth can barely survive. Such was the case with cowpea (*Vigna unguiculata*) (Fig. 6). Mycorrhizae increase the effective soil volume from which nutrients are drawn and the mycorrhizal root appears to withdraw P from solutions too dilute for effective uptake by non-mycorrhizal roots (Yost and Fox, 1979). Thus, plants which develop an effective mycorrhiza have lower external P requirements than their non-mycorrhizal counterparts (Fig. 6).

LIMING EFFECTS

Liming acid soils frequently enhances the effectiveness of P fertilizers. The external P requirement of tomatoes grown in a manganiferous soil at pH 6.3 to 7.2 was 0.2 ppm while in the pH range 5.1 to 5.7, 0.8 ppm P in solution was not sufficient for maximum yield (Jones and Fox, 1978). Under the same conditions, corn was not appreciably influenced by soil pH. These results suggest that soil pH per se did not affect P availability, but rather that low pH is associated with decreased ability of tomato roots to absorb P. This may be due to excessive uptake of Mn or Al. Under such circumstances the beneficial effects of P should be ascribed to an amendment effect of the P.

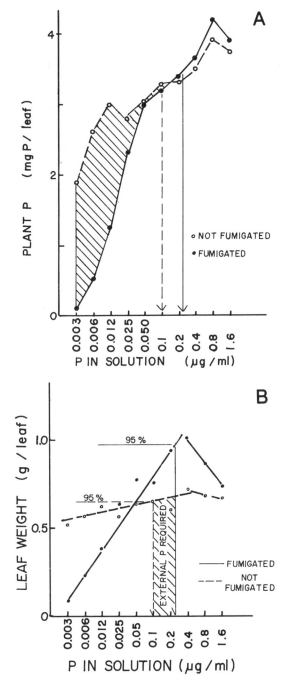

Fig. 6. Phosphorus content (A) and vigor (B) of 35 day-old cowpeas as affected by soil fumigation and soil P status. The plants grown in fumigated soil were mostly nonmycorrhizal
whereas plants growing in non-fumigated soil were heavily infected with mycorrhizae.

Transferring Information on P Requirements

The validity of the concepts presented here has been tested in several situations using a variety of crops. Fertilizer P requirements for corn and soybean, as established by field experiments, were correlated with P requirements based on P sorption curves for 12 soil series and varying levels of P fertility (Peaslee and Fox, 1978). Estimated and measured P requirements were highly correlated ($r^2 = 0.93$), with the relationship being nearly 1:1. The quantity of sorbed P in equilibrium with 0.02 ppm P in solution was a good estimate of fertilizer P required. This value is lower than had been determined in Hawaii, but is commensurate with the lower yields reported for the soils sampled.

Vander Zaag et al. (1979) plotted potato yield data from experiments performed in Bangladesh, Canada, Peru, and the U.S. (Hawaii, Idaho) against estimated levels of solution P (using P sorption curves from control samples) established by P fertilization (Fig. 7). The estimated external P requirement for potato is approximately 0.2 ppm P, a value which places potato among those crops having a high P requirement. When absolute yields were plotted against P concentration high yield potential was associated with a trend to higher external P requirements.

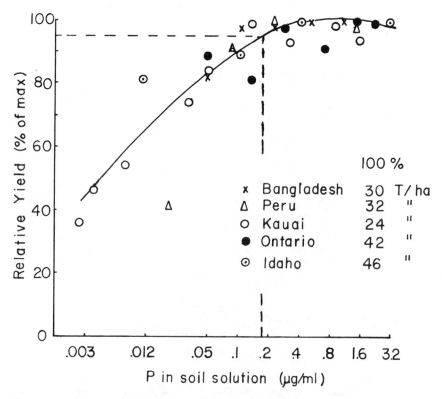

Fig. 7. A composite yield response curve for potatoes grown at five locations as a function of P concentration in solution (Vander Zaag et al., 1979).

These results demonstrate that the external P requirement concept together with standardized P sorption surves provide a rational system for recommending P fertilizer. This approach can then be expanded to create P fertilizer requirement maps. For general use such maps may be based on soil surveys if soil management practices which influence the P status of soils have been relatively uniform within a mapping unit.

SUMMARY

External P requirements of crops—that is, "concentrations of P in solutions equilibrated with soils associated with near maximum yields attainable"—are reasonably reliable for comparing the relative P requirements of crops, for placing crops into suitable ecological niches, and, if used in conjunction with P sorption curves, they provide a rational basis for predicting P fertilizer requirements. External P requirements are not greatly influenced by edaphic factors such as soil mineralogy and texture, but they seem to increase with increasing yield potential which is partially determined by day length and/or light intensity and they also increase with decreasing (low) soil temperature.

The external P requirement varies greatly among crop species but not so much among cultivars within a species. For crops like lettuce the requirement is 0.3 ppm, while for corn it is about an order of magnitude lower, and for cassave (*Manihot esculenta*) still another order of magnitude lower.

The external P requirement can be modified by the volume of soil from which roots withdraw P. Roots seem to become saturated with respect to P absorption at relatively low concentrations of P in solution—concentrations far below those obtained in unmixed fertilizer bands—which suggests that at least some mixing of P fertilizer with the soil is desirable. On the other hand, increments of P which scarcely bring the soil solution above the threshold concentration for absorption will not be efficiently utilized.

The apparent decrease in the external P requirement of some crops when soils are limed probably indicates an amendment effect of the P fertilizer in the unlimed situation resulting in diminished Al or Mn uptake by the crops.

Mycorrhizae increase the efficiency with which crops extract P from soils. They effectively decrease the external P requirement of many crop species.

Knowledge of the external P requirement of crops and P adsorption characteristics of important soils provides a rational basis for producing P fertilizer requirement maps for areas where soil survey maps are available if management factors have not obscured the soil effects.

LITERATURE CITED

1. Asher, C. J., and J. F. Loneragan. 1967. Response of plants to phosphate concentration in solution culture. I. Growth and phosphorus content. Soil Sci. 103:225–233.

2. Beckwith, R. S. 1965. Sorbed phosphate at standard supernatant concentrations as an estimate of the phosphate needs of soils. Aust. J. Exp. Agric. Anim. Husb. 5:52–58.

3. Bingham, F. T. 1949. Soil test for phosphate. Calif. Agric. 3(8):11–14.

4. Burd, J. S., and J. C. Martin. 1924. Secular and seasonal changes in the soil solution. Soil Sci. 18:151–167.

5. Cole, C. V., and S. R. Olsen. 1959. Phosphorus solubility in calcareous soils. II. Effects of exchangeable phosphorus and soil texture on phosphorus solubility. Soil Sci. Soc. Am. Proc. 23:119–121.

6. Edwards, J. H., and S. A. Barber. 1976. Phosphorus uptake rate of soybean roots as influenced by plant age, root trimming, and solution P concentration. Agron. J. 68:973–975.

7. Fox, R. L. 1974. Examples of anion and cation adsorption by soils of tropical America. Trop. Agric. (Trinidad) 51:200–210.

8. ————. 1978. Studies on phosphorus nutrition in the tropics. In C. S. Andrew and E. J. Kamprath (ed.) Mineral nutrition of legumes in tropical and subtropical soils. CSIRO.

9. ————. 1979. Comparative responses of field grown crops to phosphate concentrations in soil solutions. p. 81–106. In H. W. Mussell and R. C. Staples (ed.) Stress physiology in crop plants. Wiley-Interscience.

10. ————, and E. J. Kamprath. 1970. Phosphorus sorption isotherms for evaluating the phosphate requirement of soils. Soil Sci. Soc. Am. Proc. 34:902–907.

11. ————, and B. T. Kang. 1978. Influence of phosphorus fertilizer placement and fertilization rate on maize nutrition. Soil Sci. 125:34–50.

12. ————, R. K. Nishimoto, J. R. Thompson, and R. S. de la Pena. 1974. Comparative external phosphorus requirements of plants growing in tropical soils. 10th Int. Congr. Soil Sci. Trans. Moscow 4:232–239.

13. ————, and P. G. E. Searle. 1978. Phosphate adsorption by soils of the tropics. p. 97–119. In M. Drosdorf (ed.) Diversity of soils in the tropics. Am. Soc. of Agron.

14. Gardner, B. R., and J. P. Jones. 1973. Effects of temperature on phosphate sorption isotherms and phosphate desorption. Comm. Soil Sci. Plant Anal. 4:83–93.

15. Holford, I. C. R., and G. E. G. Mattingly. 1976. Phosphate adsorption and availability of phosphate. Plant Soil 44:377–389.

16. Jones, J. P., and J. A. Benson. 1975. Phosphate sorption isotherms for fertilizer P needs of sweet corn (Zea mays) grown on a high phosphorus fixing soil. Comm. Soil Sci. Plant Anal. 6:465–477.

17. ————, and R. L. Fox. 1978. Phosphorus nutrition of plants influenced by manganese and aluminum uptake from an Oxisol. Soil Sci. 126:230–236.

18. Juo, A. S. R., and R. L. Fox. 1977. Phosphate sorption capacity of some benchmark soils in West Africa. Soil Sci. 134:370–376.

19. Kang, B. T., and A. S. R. Juo. 1979. Balanced phosphate fertilization in humid West Africa. Phosphorus Agric. 76:75–85.

20. Leon, L. A., and W. E. Fenster. 1979. Management of phosphorus in the Andean countries of tropical Latin America. Phosphorus Agric. 76:57–73.

21. Mokurenye, U. 1979. Phosphorus needs of soils and crops of the savanna zones of Nigeria. Phosphorus Agric. 76:87–95.

22. Nishimoto, R. K., R. L. Fox, and P. E. Parvin. 1975. External and internal phosphate requirements of field grown chrysanthemums. Hort. Sci. 10:279–280.

23. ————, ————, and ————. 1977. Response of vegetable crops to phosphorus concentrations in soil solution. J. Am. Soc. Hort. Sci. 102:705–709.

24. Olsen, S. R., and F. S. Watanabe. 1963. Diffusion of phosphorus as related to soil texture and plant uptake. Soil Sci. Soc. Am. Proc. 27:648–653.

25. Parker, F. W. 1927. Soil phosphorus studies. III. Plant growth and the absorption of phosphorus from culture solutions of different phosphate concentrations. Soil Sci. 24:129–146.

26. Peaslee, D. E., and R. L. Fox. 1978. Phosphorus fertilizer requirements as estimated by phosphate sorption. Comm. Soil Sci. Plant Anal. 9:975–993.

27. Pierre, W. H., and F. W. Parker. 1927. Soil phosphorus studies. II. The concentrations of organic and inorganic phosphorus in the soil solution and soil extracts and the availability of the organic phosphorus to plants. Soil Sci. 24:423–441.

28. Tidmore, J. W. 1930. The phosphorus content of the soil solution and its relation to plant growth. Agron. J. 22:481–488.

29. Vander Zaag, P., R. L. Fox, R. de la Peña, W. M. Laughlin, A. Rhyskamp, S. Villagarcia, and D. T. Westermann. 1979. The utility of phosphate sorption curves for transferring soil management information. Trop. Agric. (Trinidad) 56:153–160.

30. Woodruff, J. R., and E. J. Kamprath. 1965. Phosphorus adsorption maximum as measured by the Langmuir isotherm and its relationship to phosphorus availability. Soil Sci. Soc. Am. Proc. 29:148–150.

31. Yost, R. S., and R. L. Fox. 1979. Contribution of mycorrhiza to the P nutrition of crops growing on an Oxisol. Agron. J. 71:903–908.

32. ————, E. J. Kamprath, E. Lobato, and G. Naderman. 1979. Phosphorus response of corn on an Oxisol as influenced by rates and placement. Soil Sci. Soc. Am. J. 43:338–343.

QUESTIONS AND ANSWERS

Q. I'm curious about the effect of liming on phosphorus especially with respect to the standard technical problem concerning the aluminum hydroxide equilibrium model. It was mentioned that liming has no effect on many soils on availability of phosphorus, and, something was said about exchangeable aluminum interacting in some way.

A. The question had to do with the effect of liming on phosphorus availability, I think. If there isn't much exchangeable aluminum in the soil system, then liming has relatively little effect on phosphate solubility. If the soil is so acid that exchangeable aluminum is a significant factor, liming will inactivate the aluminum and enhance phosphorus solubility. Agriculture in the tropics usually operates best in the pH range of about 5 to about 6.2, and in this range, liming has relatively little effect on phosphorus solubility as Dr. Sterling Olson and some of his students have found.

Q. There is some concern about phosphorus toxicity at relatively low concentrations of phosphorus. Are you pursuing this matter?

A. The Australians, particularly Drs. Asher and Loneragan, have investigated phosphorus toxicity at low concentrations of phosphorus in solution culture. We sometimes find a depression in yield when phosphorus concentrations in solution are in excess of a few parts per million or so. This may seem like a low concentration, but compared with phosphorus concentrations in most soils this is high. I do not say that this is phosphorus toxicity *per se*, but some of Dr. Loneragan's work has convinced me that he is, in fact, dealing with a phosphorus toxicity. We have not seen this in our soils to a marked degree because we carefully control the levels of phosphorus in the soil.

CHAPTER 13

Phosphorus Dissolution-Desorption in Relation to Bioavailability and Environmental Pollution[1]

D. E. PEASLEE AND R. E. PHILLIPS[2]

ABSTRACT

Managing soil fertility to assure adequate plant available P within economic and environmental constraints requires balancing the capabilities of soils to remove P from high-P-concentration solutions vs. their

---

[1] Contribution from the Dep. of Agronomy, Kentucky Agric. Exp. Stn., Lexington 40546. This paper (No. 79-3-178) is published with the approval of the Director of the Kentucky Agric. Exp. Stn. Presented before Div. S-2 of the Soil Sci. Soc. of Am. as a portion of the symposium "Chemistry in the Soil Environment," during the ASA meetings, 7 Aug. 1979, in Fort Collins, Colo.

[2] Professors of agronomy.

capabilities to release P to low-P-concentration solutions. Phosphorus supplies to plants may be sufficient at low P concentrations providing those concentrations are maintained relatively constant against plant uptake over time. The labile P/unit soil volume, the solution P concentration, the ability of a soil to replenish solution P, and the P replenishment rate, all interact to control the overall P supply to plants. There can be adjustments and balances among these factors to achieve adequate supplies within a particular soil.

Replenishment capabilities, generally defined as buffering "capacities" or "powers," describe the relationship between solid and solution phases in soils, and are extremely important in plant nutrition, as well as having impacts on environmental pollution and computer simulation modeling. Despite the broad implications of phosphate buffering power, numerous methods have been used for measuring this index, without benefit of critical studies that would provide a basis for selecting the most appropriate method.

In this review article, several methods for evaluating phosphate buffer powers have been compared by evaluating their effect on calculated diffusive fluxes. As a general technique, buffer powers determined from desorption data were more valid than those determined from adsorption data. Phosphorus uptake rates by plant roots and P release rates from soils as reported in the literature agree qualitatively assuming that 1 to 5% of the soil volume supplies P to plants. It is important that the adjustments among buffering properties, solution concentrations, and release rates be understood and utilized in maximizing P supply and minimizing potential environmental problems.

INTRODUCTION

Phosphate apparently is adsorbed from dilute solutions onto high affinity sites, then, as these become increasingly saturated or as solution concentrations become greater, P is adsorbed onto lower affinity sites and/or precipitated as relatively soluble compounds. Since adsorption is relatively rapid and precipitation of P compounds is relatively slow, solutions from fertilizer particles in the field and from most test tube solutions in the laboratory result in "concentration overrun," as far as permitting the soil to react with P in classical equilibrium terms. Therefore, soil scientists have been forced to accept the P chemistry in soils as being dynamic at worst, or as exhibiting "heterogeneous equilibria" at best.

In practical terms, transfer of P from solid to solution phases in a continuing depletion system (such as a soil-plant system or under leaching conditions) may be considered as dissolution of precipitated P or as P desorption. In reality, it is probably a combination of both processes occurring simultaneously. Larsen (26) suggested that in neutral and calcareous soils distinctions between adsorbed and precipitated soil P was immaterial since the concepts simply arise from different views of the same system. Release of P from slightly acid soils (39) during cyclic 15-min extractions is shown in Fig. 1. Release rates and intraparticle diffusion should have had an effect. Under these conditions, there was a complex

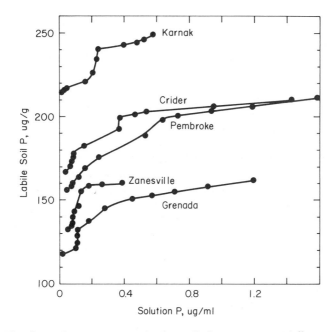

Fig. 1. Phosphorus desorption curves for five soils that were sequentially extracted with 0.001N CaCl$_2$ for 350, 15-min cycles (38).

pattern of release that indicated changes in sources of soil P during extraction. Solubility criteria applied to the data indicate initial undersaturation with respect to octacalcium phosphate and finally, during subsequent extractions, undersaturation with respect to hydroxyapatites (39). Continuous leaching of soil columns by gravity (14) did not delineate marked changes in release patterns within soils. Two types of release patterns among soils were observed. Soils exhibiting a continuous decline in concentration were assumed to contain adsorbed P, whereas those soils having more stable concentrations in their leachate were assumed to contain soluble P compounds. Regardless of chemical identities of soil P, there are several release properties that control plant root supply. These properties are:

a. concentration of labile P in soils
b. concentration of P in soil solution
c. relative constancy of solution P concentration during P depletion
d. rate of solid-to-solution transfer.

NUTRIENT SUPPLYING PROPERTIES

Buffering Characteristics

"Capacity factor," "phosphate potential," "buffer capacity," and "buffer power" are terms used to quantify the tendency of a soil to resist a change in ion concentrations in the solution phase (44, 50, 31, 19). A buffer property has intensive as well as extensive properties (31) and cannot correctly be designated as a capacity property. Buffer power will be the term used here and will be defined as the average change in labile soil P (ΔC) in relation to a change in solution P (ΔC_l) with units ($\mu g/cm^3$ soil)/ ($\mu g/cm^3$ soln).

Practical and mathematical application of buffer properties have been hampered by two characteristics of buffering factors. In a 1971 review on soil solution properties, Khasawneh (24) illustrated the continuous variation of buffer power for nutrient ions as the magnitude of q and I was varied. The terms q and I are analgous to labile C and solution C, respectively, as defined above. Differential buffer power, dq/dI, (24) permits evaluating the non-linear relationship between soil and solution distributions of nutrient ions. Figure 2 shows a hypothetical phosphate buffer curve to illustrate buffer power and differential buffer power. Recognizing the lack of buffer power constancy of soils, especially for P, recently has prompted the proposal of a "maximum buffer capacity" (18). Divergence between soil adsorption and desorption isotherms also causes problems in evaluating buffer properties as illustrated for Grenada and Zanesville silt loams (Glossic and Typic fragiudalfs, respectively). These samples were similar in texture, pH, organic matter, and initial extractable P (5). After equilibrating soil samples with serially increasing solution P concentrations, they were immediately desorbed without drying (J. C. Ballaux. 1972. Adsorption and desorption of P in five Kentucky soils. Ph.D. thesis. Univ. of Kentucky, Lexington). Substantial differences are evident in P adsorption-desorption isotherms shown in Fig. 3.

Causes of hysteresis may be: a) bidentate bond formation by phosphate ions after sorption (22); b) diffusion into crystalline or amorphous materials (51); c) recrystallization of insoluble compounds (28, 43); or d)

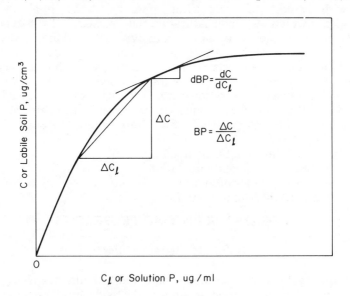

Fig. 2. Hypothetical curve illustrating buffer and differential buffer characteristics of soils.

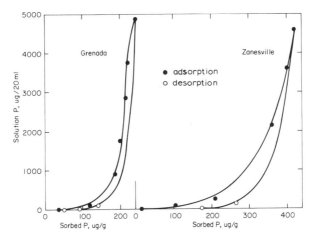

Fig. 3. Hysteresis effects in P sorption and desorption of Grenada and Zanesville silt loam soils.

replacement of silica in surface tetrahedral positions in clay minerals (28). Combinations of these processes, or lack of complete equilibrium also may cause hysteresis. Nonconstancy of $\Delta C/\Delta C_l$ for a given soil system, delineation of apparent discreet pools of labile P by desorption techniques that are not discernible by adsorption curves, as well as the phenomena of hysteresis, have raised questions concerning the most correct method for measuring phosphate buffering properties of soils. Clearly, net transfer tendencies of phosphate from soil to solution is the reaction requiring characterization. This dilemma is indicated by the multitude of techniques reported in the literature: [31]P adsorption (19); ^{32}P adsorption exchange (34, 35) ^{31}P desorption by resin (8); ^{31}P desorption with dilute $CaCl_2$ (10). With the exception of evaluations for "maximum buffer capacity" and "equilibrium buffer capacity," both determined from adsorption data, there have been few studies where techniques for measuring buffer powers were compared on the basis of relative abilities of soils to release P. Buffer powers were calculated on two similar soils using data from cyclic desorption (39); from ^{31}P adsorption in $0.001N$ $CaCL_2$[3]; from ^{31}P desorption by anion resin (5); and from ^{32}P exchangeable P in samples[3] pre-treated with increasing rates of KH_2PO_4. Desorption by resin and isotopically exchangeable P frequently are determined on samples pre-treated with P. With these techniques, C tends to increase linearly over rather broad ranges as C_l increases, and will tend, therefore, to be less sensitive to effects of high affinity sites or to other discrete pools of labile P. To the extent possible, the buffer power was determined for all methods at similar values of C_l ($\cong 0.05$ $\mu g/ml$). Two to three-fold variation in the order of magnitude of buffer powers within soils as shown in Table 1 is a disturbing trend. A more serious discrepancy, though is the trend toward larger estimates of buffer power for the Zanesville compared with the Grenada soil from adsorption data, whereas the re-

[3] D. E. Peaslee, unpublished data.

Table 1. Phosphate buffer powers of Grenada and Zanesville soils as determined by four methods.

Soil	Buffer power			
	Sequential desorption	Adsorption	^{32}P exchange	Resin desorption
	(μg/cm^3 soil)/(μg/cm^3 solution)			
Grenada	677	568	271	615
Zanesville	384	906	250	490

verse trend was obtained from desorption data. Buffering based on ^{32}P exchange data was only slightly greater for Grenada, than for Zanesville, but the ranking was the same from ^{32}P and from desorption data.

A relatively large number of high affinity sites were apparently present in the Zanesville soil resulting in strong buffering tendency based on adsorption data (Fig. 4). During desorption in the range of 0.05–0.1 μg/ml, however, concentrations in extracts were more stable for the Grenada than for the Zanesville, resulting in a larger buffer factor for the Grenada based on desorption data (Fig. 5). Initial labile P was 2 and 1 μg/g for the Grenada and Zanesville soils, respectively. To illustrate what effect differences of this magnitude would have on calculated diffusion to roots, buffer powers from cyclic desorption and from adsorption were used in diffusive flux equations (32, 33) while holding all other parameters constant.

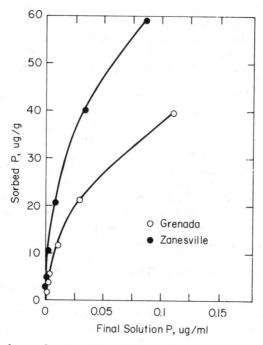

Fig. 4. Phosphorus adsorption curves for Grenada and Zanesville soils used in estimating differential phosphate buffering characteristics. Estimated curve fit.

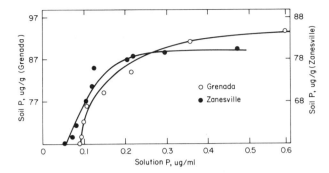

Fig. 5. Phosphorus desorption curves for Grenada and Zanesville soils used in estimating differential phosphate buffering characteristics. Estimated curve fit.

P Flux Equations and Buffer Power

Equation [20] of Olsen and Kemper (33) given below as Eq. [1] was used to calculate Q, the total amount of P absorbed by the root per unit surface area in time, t

$$Q = aB\,(C_0 - C_r)\left[\frac{2}{(\pi)^{1/2}}\,T^{1/2} + \frac{1}{2}\,T - \frac{1}{6}\,\pi T^{3/2} + \frac{1}{16}T^2 \ldots\right] \quad [1]$$

where

Q = amount of P absorbed per unit area of root surface, $\mu g/cm^2$ of root

B = b + θ; b = capacity factor and θ = volumetric water content

C_0 = concentration of P in soil solution, $\mu g/ml$

C_r = concentration of P in soil solution at the root surface, $\mu g/ml$

T = $D_p t/a^2 B$; D_p = porous diffusion coefficient, cm^2/sec; t = time in sec; a = radius of root, cm.

Q was divided by t to give the average flux, $\bar{q}$, the amount of P absorbed by the root per unit surface area per unit time. Equation [1] assumes a constant P concentration at the root surface. The $\bar{q}$ value thus calculated was then used in Eq. [21] of Olsen & Kemper (33) given below as Eq. [2] to calculate C_r, the P concentration in solution as a function of time and distance (r − a) from the root surface:

$$C_0 - C_r = \frac{2\bar{q}}{D_p}\left(\frac{a}{r}\right)^{1/2}(D_p t/B)^{1/2}\left[\operatorname{ierfc}\left(\frac{r-a}{2(D_p t/B)^{1/2}}\right)\right.$$
$$\left. - \left(\frac{3r+a}{4ar}\right)(D_p t/B)^{1/2}\,i^2\operatorname{erfc}\left(\frac{r-a}{2(D_p t/B)^{1/2}}\right) + \ldots\right] \quad [2]$$

Equation [2] assumes a constant rate of uptake per unit area of root surface. The values of D_p and "a" assumed in all calculations were 4.3 × 10⁻⁷ cm²/sec and 0.05 cm, respectively.

There is some question about using the procedure outlined above but there is some C_r value which would give the correct $\bar{q}$ value. It is not known how low the P concentration will become at the root surface. It probably will approach zero when the plant is P deficient, and it may approach C_0 when the plant is well supplied with P. If C_r is selected too close to C_0, then the $\bar{q}$ value calculated from Eq. [1] would be too small, but if C_r is allowed to equal zero, then the calculated $\bar{q}$ value may be too large.

Calculated Diffusion Fluxes and Depletion Zones

The Grenada soil exhibited only slight hysteresis (Fig. 3) and calculated fluxes were similar, whether desorption or adsorption buffer powers were used (Table 2). For the Zanesville soil, fluxes were greater when calculations were based on adsorption buffer. Phosphorus released during cyclic extraction with $0.001N$ $CaCl_2$ or extraction by resin agreed qualitatively with flux calculations based on desorption, but were inversely related to fluxes based on adsorption data. Effects of concentration on calculated fluxes were compared for the Grenada soil by holding parameters in Eq. [1] and [2] constant and increasing C_0 by the same relative amount as the desorption buffer factor increased between the Zanesville and Grenada soils. Increasing the buffer power and C_0 by a factor of 1.76 increased the fluxes (Table 2) by factors of 1.51 and 2.14 for buffer and concentration effects, respectively.

Depletion zones at $t = 10$ days around a hypothetical root were calculated from Eq. [2], first calculating concentrations in solution as a function of distance from the root surface and then calculating C from Eq. [3]:

$$C = A + (\text{Buffer power})\, C_l \qquad [3]$$

When calculations were based on desorption data and $C_0 = 0.05$ $\mu g/ml$ (Fig. 6), depletion as a function of distance was greater for the Grenada soil than for the Zanesville soil to a distance of about 0.035 cm. From 0.035 to 0.065 cm (at the point where $C_r \cong C_0$), depletion was essentially identical for the two soils. Figure 6 also illustrates the effect of

Table 2. Calculated fluxes of P at root surfaces based on adsorption and desorption buffer properties of Grenada and Zanesville soils and quantities of extractable P.

| Soil | Calculated flux from | | Measured P release | |
	Desorption buffer	Adsorption buffer	Sequential extraction	Resin extraction§
	— $\mu g/cm^2$ sec $\times 10^7$ —		— $\mu g/g$ —	
Grenada†	5.6	5.2	47	4
Zanesville†	3.7	6.5	28	34
Grenada‡	12.0	--	--	--

† $C_0 = 0.05\,\mu g/ml$.
‡ $C_0 = 0.088\,\mu g/ml$.
§ Interpolated at $C_0 = 0.07\,\mu g/ml$.

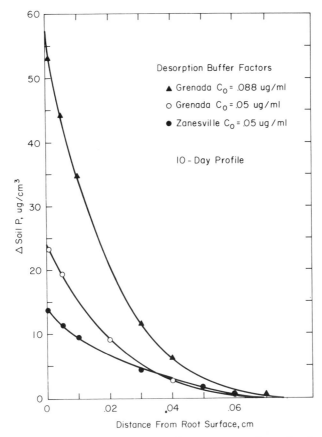

Fig. 6. Calculated soil P utilization profiles adjacent to root surface using desorption data for Grenada and Zanesville soils.

increasing C_0 from 0.05 to 0.088 μg/ml for the Grenada soil. When C_0 = 0.088 μg/ml, depletion was greater from the root surface to a distance of > 0.06 cm. Depletion profiles from adsorption buffer factors (Fig. 7) were reversed for these two soils compared with those obtained by using desorption buffer factors.

For some soils at least, phosphate buffer properties are more correctly evaluated by desorption measurements. Nye and coworkers (29, 30, 49) have suggested that desorption is the superior technique for estimating buffer power and its effect on nutrient uptake by plant roots. Brewster et al. (8) measured adsorption and desorption isotherms for a "well equilibrated" calcareous sandy loam. They concluded that desorption by anion resin and adsorption conducted at low soil-to-solution ratios were satisfactory for estimating buffer power of that soil, though numerically, values were different between the two techniques.

Relatively little data are available illustrating unconfounded effects of buffer characteristics of soils on plant uptake of P. Usually, concentration in soil solution, diffusion coefficients, water content, or other factors

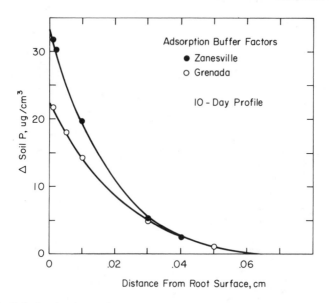

Fig. 7. Calculated soil P utilization profiles adjacent to root surface using adsorption data for
Grenada and Zanesville soils.

differ among soils, thereby confounding the effects of buffering properties. Data from greenhouse experiments have shown that buffer properties
and concentration properties (4, 15, 18, 19) were important in plant P uptake. Holford and Mattingly (18, 19) proposed the two-surface Langmuir
equation as a model for P adsorption by soils and the use of the equation
constants to calculate a "maximum buffer capacity" (MBC) as measured
by adsorption techniques. With eight calcareous soils from the same soil
series, each at three levels of P, and ranging in pH from 7.5 to 8.0, they
concluded that MBC was a more important index of plant available P
than was "equilibrium buffer capacity" (EBC) as measured by P adsorption. However, more variance in P uptake by plants was accounted for by
initial concentration of P in solution (C_0), and quantity of P on high-
affinity sites than by C_0 in combination with either MBC or ECB (R^2 =
0.90, 0.79, and 0.30, respectively).

Phosphate buffering characteristics will become increasingly important as modeling of nutrient movement to plant roots expands as a tool
in soil fertility and crop management, and as potential loading of soil with
P and modeling of its movement with soil water are used in environmental research. Methodology in characterizing buffer properties of soils
should be selected logically, and within limits of practicality, conditions
of measurement should be influenced by the objectives. For example,
phosphate load potentials for environmental purposes should probably be
determined by adsorption techniques at relatively high initial P concentrations, large solution-to-soil ratios, and short contact times, whereas
capabilities of soils to supply P to plants should probably be determined
by desorption techniques at relatively low concentrations, small solution-
to-soil ratios, and short contact times.

CONCENTRATIONS IN SOIL SOLUTION

Absorption rates at root surfaces, and the physical-chemical environment through which P diffuses to the root determines the solution concentration, C_0, that will produce the required flux. Inseparable from this is the proficiency with which the solution concentration is maintained at the diffusion boundary, as indicated by the buffer power. Asher and Loneragan (3) demonstrated that adequate P can be accumulated by plants even at very low solution concentrations in continuous flowing nutrient solutions, where buffer powers approach infinity. In soil systems where buffer powers are in the range of 150 to 1,000, and where diffusion may be limiting, solution concentrations that are necessary for adequate P supply are often greater than those reported for nutrient solutions.

Larsen (26, 27), Khasawneh (24), Adams (1) and Nye (29) have thoroughly discussed methods of characterizing soil solutions, calculating activities of ion species, and pros and cons of using ion activities, so these issues will not be discussed here. Collectively, the general conclusions were that ion activities often are significant refinements over concentrations where systems are well defined, that solution properties should be characterized at low solution-to-soil ratios, and that controlling partial pressures of CO_2 is critical during equilibration of soils containing significant quantities of basic calcium phosphates.

Saturation extracts and displaced pore solutions are normally measured with deionized H_2O, whereas extractions at solution-to-soil ratios of > 1 normally are made with dilute electrolytes to aid flocculation. Since concentration of P in soil extracts is often influenced by Ca concentrations, soils have frequently been extracted with solutions of 0.001 to 0.01M $CaCl_2$.

Critical concentrations for soil solution P reported in the literature may vary then according to plant uptake conditions, especially time and rooting density, extraction method, and buffer factor. In 0.01M $CaCl_2$ extracts, estimated requirements were 0.02 μg P/ml (40), 0.05 μg/ml (25), and 0.37 μg/ml (46). The latter value was estimated by extracting 20 or 40 g soil with 50 ml of solutions for a "few seconds," then extrapolating to zero dilution; from the same experiment (46), P requirement from displaced solutions was 0.68 μg/ml. From a study of 24 soil samples of the same series, Holford and Mattingly (19) estimated critical solution concentrations in 0.02M KCl; requirements decreased from 0.095 to 0.02 μg/ml as "maximum buffer capacities" increased from about 2,500 to about 9,000 ml/g of soil.

From data cited above, critical concentrations appeared to be lower when soil-to-solution ratios were lower. Concentrations of solution P at equilibrium are generally considered to be independent of soil-to-solution ratios, but Larsen and Widdowson (27) identified CO_2 effects as the cause of increasing solution P with increasing soil-to-solution ratios. As determined by standard techniques, P in displaced pore solutions represents that quantity which will equilibrate with a small water volume after an initial, extended contact time. Depending on the technique used, then, measurements with displaced pore solutions may tend to over estimate solution P. Peaslee and Ballaux (39) measured relatively constant P concentrations of about 0.08 to 0.1 μg/ml in 15-min cyclic extractions of 5 g soil with 3.5 ml of 0.001N CaCl$_2$. Initial concentrations in extracts from these highly fertilized soils ranged initially up to 1 μg/ml, but rapidly decreased after a few extractions to much lower and more stable levels.

Diversity of experimental conditions and techniques prevents selection of a critical concentration or range from the literature. However, at solution-to-soil ratios between 1:1 and 5:1, and for neutral to slightly acid soils with medium texture, critical concentrations ranged from 0.05 to 0.1 μg/ml. These concentrations are consistent with frequency distribution of concentrations in saturation extracts of cultivated, midwestern soils in which most extracts contained 0.04 to 0.06 μg/ml (6).

RATES OF PHOSPHATE RELEASE

Differences in kinetics of P release among soils are usually confounded by other soil differences such as concentrations in the solid phase or in the soil solution. Rates of release differ widely among soils and within soils over time.

Subject to variations in root surface area/soil volume, the mean rate of plant removal of P per unit of soil should provide a reference point for required rates of release from soil. Khasawneh and Copeland (25) grew plants for a period of 8 weeks in a highly fertilized, fine sandy loam soil; from their data, the mean uptake rate was about 1×10^{-6} μg/g/sec. Similar rates were observed over a 14-day period by other researchers with a sandy loam soil (46). Our calculations from short-term uptake studies on clay soils (38) gave rates of 1×10^{-5} μg/g/sec. For medium textured soils, rates were about 5×10^{-7} to 55×10^{-7} μg/g/sec for low and high P levels, respectively, over a 125-day period (Sayan Singholka. 1970. Adsorption and bonding energy indexes of P in 15 Kentucky soils. M.S. thesis. Univ. of Kentucky, Lexington). A nominal rate of P removal from soil appears to be 50×10^{-7} μg/g/sec.

Since roots exploit only about 1 to 5% of the soil, the capability of the soil to supply P should be 20 to 100-fold greater than plant requirements on a unit soil basis. The approximation agrees well with data in Table 3

Table 3. Average rates of transfer of P from solid to solution phases.

Extraction method	Transfer rate	Data source
	P, μg/g-sec $\times 10^5$	
Resin	28	2
Resin	32	48
Resin	98	38
Resin	3	38
Resin	17	5
Resin	58	5
Resin	42	11
Resin	190	13
Leaching	18	5
Cyclic extraction	29	5

which shows P extracted by anion or anion-cation resins from soil suspensions (2 to 16 hours) and from leaching or from extraction experiments. The data cited for Amer et al. (2) was for the slow release phase and data cited for Peaslee and Ballaux (39) was obtained during the intermediate extraction phase (extraction cycles 26 to 150).

ENVIRONMENTAL IMPLICATIONS

Solution-solid phase relations for P in soils has an obvious impact on environmental science. Harter and Foster (17) have recently based a computer simulation model for P movement through soils on P adsorption properties. The potential for a soil to receive P at a waste disposal site depends on several complex factors, one of which is the ability of the soil to remove P from solution. For these purposes phosphate adsorption buffer factors should be better indexes than desorption buffer factors. At some point in time, however, the soil may cease being a waste disposal area. Desorption buffer properties may then become important factors in determining how much, and under what conditions, P will be released from the soil. These properties should be determined before using disposal sites.

Concentrations of P in most soil solutions (6, 39) should not lead to environmental problems. However, some soil solutions have P concentrations of > 0.1 $\mu g/ml$ (47) due to high P contents of parent material. Some sands in agricultural production require special precautions to avoid pollution because of low phosphate buffering properties and extensive use of irrigation. Even under limited rainfall ($\cong 56$ cm), 81% of surface-applied fertilizer P moved below the 10-cm depth in Australia (36); distribution by depth was 22, 17, and 22% of surface applied P in the 0 to 10, 10 to 30, and 30 to 60-cm depths, respectively. Some acid organic soils also may permit leaching of P (12). Two somewhat unique aspects of P behavior in waste disposal problems are the effect of highly labile carbon on redox potentials in soils (23), and the extra dimension caused by the presence of organic P compounds (16), and soluble organic matter in the presence of inorganic P (45). When energy is available and conditions are favorable for high levels of biological activity, reduced soil conditions often occur in local zones or micro-sites even under "aerobic" conditions. As wetness increases to the point of being waterlogged or flooded, the portion of reduced soil may approach totality. Flooding of soils often increases soil test indexes for P available to plants (41), and this has been attributed primarily to reduction of Fe in the soil, thereby solubilizing $FePO_4$ as well as occluded and sorbed P associated with ferric oxyhydroxides. The extent of this reaction is affected by pH, P status and reducible Fe in the soil. Patrick and co-workers (21, 23, 41) have recently measured release of soil P and sorption of added P in several soils at various combinations of redox potential and pH. Solubility of soil P was greatly increased by reducing conditions, especially at pH 5, as a result of dissolution of iron compounds without forming new sorption surfaces such as $Fe(OH)_3$ gels. Strongly reduced soils at pH 6.5 had slightly increased soluble P, and large increases in sorption capacities and buffer properties; these changes were attributed to formation of $Fe(OH)_3$ at pH 6.5, which served as sorption surfaces. In general, their data clearly illustrate that soils having received high levels of P application and having at least nominal amounts of labile Fe may release rather high concentrations of P to solution under reducing conditions. At high concentrations of P, sorption capacities (and sorbed P) were greater under reduced than under normal conditions. The important phenomena, however, is the increase in solubility of soil P under reducing conditions; e.g. solution concentrations for a soil with nominal P levels increased from 0.06 to 1.64 $\mu g/ml$ for normal and reduced conditions, respectively (23).

Availability of organic P to plants and the behavior of organic P in soils are worthy as separate topics in their own right, so they cannot receive just attention here. Existence of inositol phosphates in soils has been recognized for decades (7, 9). Evidence has been presented to indicate the presence of significant amounts of organic P in soils (52), and higher concentrations of organic P compared with inorganic P ($\cong 0.6$ to 0.07 $\mu g/ml$, respectively) in leachates from soils after decomposition of plant material (16) shows a need for closer monitoring of organic P in systems with high biological activity. This is supported by data from a leaching study using a range of organic P compounds (42). Glycerophosphate leached to greater depths than inorganic P and underwent rapid hydrolysis.

SUMMARY

In summary, information is being collected that characterizes soils in terms of their role in behavior of P and eventually its availability to plants. The concentration in solution, C_l, the tendency to be transferred from solid to solution phases, $\Delta C/\Delta C_l$ and the rate of this transfer are subject to large differences depending on how they are measured. In view of the far-reaching impact these parameters are likely to have on agronomic and environmental policies, the most accurate methods for measuring these parameters for a particular set of objectives and soil types should be determined.

LITERATURE CITED

1. Adams, Fred. 1971. Ionic concentrations and activities in soil solutions. Soil Sci. Soc. Am. Proc. 35:420–426.
2. Amer, F., D. R. Bouldin, C. A. Black, and F. R. Duke. 1955. Characterization of soil phosphorus by anion exchange resin adsorption and P-32 equilibration. Plant Soil VI: 391.
3. Asher, C. J., and J. F. Loneragan. 1967. Response of plants to phosphate concentration in solution culture. I. Growth and phosphorus content. Soil Sci. 103:311–318.
4. Baldovinos, F., and G. W. Thomas. 1967. The effect of soil clay content on phosphorus uptake. Soil Sci. Soc. Am. Proc. 31:680–682.
5. Ballaux, J. C., and D. E. Peaslee. 1975. Relationships between sorption and desorption of phosphorus by soils. Soil Sci. Soc. Am. Proc. 31:680–682.
6. Barber, S. A., J. M. Walker, and E. H. Vasey. 1962. Principles of ion movement through the soil to the plant root. Int. Soc. Soil Sci. Trans. Joint Meeting Comm. IV & V (New Zealand) 1962:121–125.

7. Bower, C. A. 1949. Studies on the forms and availability of soil organic phosphorus. Iowa Agric. Exp. Stn. Res. Bull. 362.

8. Brewster, J. L., A. N. Gancheva, and P. H. Nye. 1975. The determination of desorption isotherms for soil phosphate using low volumes of solution and an anion exchange resin. J. Soil Sci. 26:364–377.

9. Cosgrove, D. J. 1963. The chemical nature of soil organic phosphorus. I. Inositol phosphates. Aust. J. Soil Res. 1:203–214.

10. Drew, M. C., and P. H. Nye. 1970. The supply of nutrient ions by diffusion to plant roots in soil. III. Uptake of phosphate by roots of onion, leek, and rye-grass. Plant Soil 33:545–563.

11. El-Nennah, M. 1978. Phosphorus in soil extracted with anion exchange resin. I. Time-dissolution relationship. Plant Soil 49:647–651.

12. Fox, R. L., and E. J. Kemprath. 1971. Adsorption and leaching of P in acid organic soils and high organic matter sand. Soil Sci. Soc. Am. Proc. 35:154–156.

13. Fried, Maurice, C. E. Hagen, J. F. Saiz Del Rio, and J. E. Leggett. 1957. Kinetics of phosphate uptake in the soil-plant system. Soil Sci. 84:427–437.

14. ————, and R. E. Shapiro. 1956. Phosphate supply pattern of various soils. Soil Sci. Soc. Am. Proc. 20:471–475.

15. Gunary, D., and C. D. Sutton. 1967. Soil factors affecting plant uptake of phosphate. J. Soil Sci. 18:167–173.

16. Hannapel, R. J., W. H. Fuller, Shirley Bosma, and J. S. Bullock. 1964. Phosphorus movement in calcareous soil. I. Predominance of organic forms of phosphorus in phosphorus movement. Soil Sci. 97:350–357.

17. Harter, R. D., and B. B. Foster. 1976. Computer simulation of phosphorus movement through soils. Soil Sci. Soc. Am. J. 40:239–242.

18. Holford, I. C. R., and G. E. G. Mattingly. 1976. A model for the behavior of labile phosphate in soil. Plant Soil 44:219–229.

19. ————, and ————. 1976. Phosphate adsorption and plant availability of phosphate. Plant Soil 44:337–389.

20. ————. 1977. Soil properties related to phosphate buffering in calcareous soils. Soil Sci. Plant Anal. 8:125–137.

21. ————, and W. H. Patrick, Jr. 1979. Effects of reduction and pH changes on phosphate sorption and mobility in an acid soil. Soil Sci. Am. J. 43:292–297.

22. Kafkafi, U., A. M. Posner, and J. P. Quirk. 1967. Desorption of phosphate from kaolinite. Soil Sci. Soc. Am. Proc. 31:348–353.

23. Khalid, R. A., W. H. Patrick, Jr., and R. D. DeLaune. 1977. Phosphorus sorption characteristics of flooded soils. Soil Sci. Soc. Am. J. 41:305–310.

24. Khasawneh, F. E. 1971. Solution ion activity and plant growth. Soil Sci. Soc. Am. Proc. 35:425–436.

25. ————, and J. P. Copeland. 1973. Cotton root growth and uptake of nutrients: Relation of phosphorus uptake to quantity, intensity, and buffering capacity. Soil Sci. Soc. Am. Proc. 37:250–254.

26. Larsen, Sigurd. 1967. Soil phosphorus. Adv. Agron. 19:151–210.

27. ————, and A. E. Widdowson. 1964. Effect of soil/solution ratio on determining the chemical potentials of phosphate ions in soil solutions. Nature 203.942.

28. Low, P. F., and C. A. Black. 1950. Reactions of phosphate with kaolinite. Soil Sci. 70:273–290.

29. Nye, P. H. 1966. The effect of nutrient intensity and buffering power of a soil, and the absorbing power, size and roothairs of a root, on nutrient absorption by diffusion. Plant Soil 25:81–105.

30. ————. 1968. The use of exchange isotherms to determine diffusion coefficients in soil. 9th Int. Congr. Soil Sci. Soc. Trans. Adelaide 1:117–126.

31. ————, and P. B. Tinker. 1977. Local movement of solutes in soils. p. 69–91. In Solute movement in the soil-root system. Univ. of California Press, Berkeley.

32. Olsen, S. R., W. D. Kemper, and R. D. Jackson. 1962. Phosphate diffusion to plant roots. Soil Sci. Soc. Am. Proc. 26:224–227.

33. ————, and ————. 1968. Movement of nutrients to plant roots. Adv. Agron. 20: 91–151.

34. ————, and S. Watanabe. 1963. Diffusive supply of phosphorus in relation to soil textural variations. Soil Sci. 110:318–327.

35. ————, and ————. 1963. Diffusion of phosphorus as related to soil texture and plant uptake. Soil Sci. Soc. Am. Proc. 27:648–653.

36. Ozanne, P. G., and D. J. Kirton. 1961. The loss of phosphorus from sandy soils. Aust. J. Agric. Res. 12:309–423.

37. Patrick, W. H., Jr., and R. A. Khalid. 1977. Phosphate release and sorption by soils and sediments: Effect of aerobic and anaerobic conditions. Science 186:53–55.

38. Peaslee, D. E. 1969. Indexes of availability of K, Mg, Ca, and P in Connecticut soils. Agron. J. 61:330–331.

39. ————, and J. C. Ballaux. 1977. Short-term replenishment of soils solution phosphorus. Soil Sci. Soc. Am. J. 41:529–531.

40. ————, and R. L. Fox. 1978. Phosphorus fertilizer requirements as estimated by phosphate sorption. Soil Sci. Plant Anal. 9:975–993.

41. Redman, F. H., and W. H. Patrick, Jr. 1965. Effect of submergence on several biological and chemical soil properties. Bull. 592. Louisiana State Univ. Agric. Exp. Stn., Baton Rouge.

42. Rolston, D. E., R. S. Rauschkolb, and D. R. Hoffman. 1975. Infiltration of organic phosphate compounds in soil. Soil Sci. Soc. Am. Proc. 39:1,089–1,094.

43. Russell, R. S., J. B. Rickson, and S. N. Adams. 1954. Isotopic equilibria between phosphates in soils and their significance in the assessment of fertility by tracer methods. J. Soil Soc. 5:85–105.

44. Schofield, R. K. 1955. Can a precise meaning be given to "available" soil phosphorus? Soils Fert. 18:373–375.

45. Singh, B. B., and J. J. Jones. 1976. Phosphorus sorption and desorption characteristics of soil as affected by organic residues. Soil Sci. Soc. Am. J. 40:389–394.

46. Saltanpour, R. N., Fred Adams, and A. C. Bennett. 1974. Soil phosphorus availability as measured by displaced soil solutions, calcium-chloride extracts, dilute-acid extracts, and labile phosphorus. Soil Sci. Soc. Am. Proc. 38:225–228.

47. Thomas, G. W., and J. D. Crutchfield. 1974. Nitrate-nitrogen and phosphorus contents of streams draining small agricultural watersheds in Kentucky. J. Environ. Qual. 3:46–49.

48. Vaidyanathan, L. V., and O. Talibudeen. 1968. Rate controlling processes in the release of soil phosphate. J. Soil Sci. 19:342–353.

49. ————, and P. H. Nye. 1972. The measurement and mechanism of ion diffusion in soils. VI. The effect of concentration and moisture content on the diffusion of soil phosphate against chloride ion. J. Soil Sci. 21:15–27.

50. White, R. E., and P. H. T. Beckett. 1964. Studies on the phosphate potentials of soils. I. The measurement of phosphate potential. Plant Soil 20:1–16.

51. ————, and A. W. Taylor. 1977. Reactions of soluble phosphate with acid soils: The interpretation of adsorption-desorption isotherms. J. Soil Sci. 28:314–328.

52. Wild, A., and O. L. Oke. 1966. Organic phosphate compounds in calcium chloride extracts of soils: Identification and availability to plants. J. Soil Sci. 356–371.

QUESTIONS AND ANSWERS

Q. You used phosphorus as your illustration, and I agree that in this case there is a big difference between adsorption and desorption. Determining buffer power by desorption would seem to be the better way of doing it. Now in the case of exchangeable cations, would you consider that there probably is not much diffeence between adsorption and desorption isotherms?

A. What I have seen in the literature for potassium, at least, indicated that in the case of mica-type clay minerals, there was a difference between the adsorption and desorption estimates of the buffer capacity. I don't think there is as severe a problem for potassium as there is for phosphorus on whether we do a desorption or adsorption measurement, but in some special cases there may be a problem.

Q. a. Would you expect differences in adsorption versus desorption if you were doing this on an equilibrium basis where very small amounts were being sorbed or desorbed? Would you not expect them to be the same there, and aren't you simply using a different point on the curve or degree of saturation which would account for the difference?

Q. b. If I understood it right, the question is whether or not he'd expect the same difference between dissolution and desorption versus adsorption if you use small increments. . ., is that right?

Q. c. Yes. The question that comes to mind, of course, is whether you're dealing with the same adsorption site, whether you had exceeded certain ones that might be different for adsorption versus desorption and whether you'd in fact be able to get the same value whether it went on or came off if you were using, say, equilibrium procedures with just very small amounts going on or coming off.

A. Right. There has been some data presented by Holford and Mattingly of England, on calcareous soils. They studied 24 samples, eight sampling sites, and three different phosphorus levels on a single soil series. They evaluated buffer capacities at equal solution concentrations. Their data would convince you that the key point is equilibrium at a particular solution concentration. This is one reason why I tried to point out that we evaluated these at 0.074 micrograms per ml. Now I am convinced from our data that equilibrium is a key issue and that the total phosphorus level and the solution phosphorus level are both critical. What we attempted to do was to hold these constant as best we could. Now where were we in equilibrium? I'm not sure. But neither do I consider the soil plant system in field crop production as being in equilibrium. In our studies, we hope that we were at least reasonably close to equilibrium. I think you can get conditions at which you can do up and down these adsorption-desorption curves without hysteresis. The Grenada soil tended to do this, the Zanesville did not. There may be a reason for that. Dr. Nye in England has proposed that desorption is much more valid than adsorption in reflecting nutrient availability. I think there are controversies on the subject, but from our data, we would have to choose desorption as the process to follow in evaluating bioavailability.

Q. a. I've only observed that one case, one soil in which there is not a hysteresis loop in adsorption-desorption curves. Usually the hysteresis loop is quite pronounced, and if you desorb intensively the hysteresis is even more pronounced.

Q. b. In other words, it's fairly well agreed that there is a hysteresis between adsorption and desorption, so the question I would have is, do
we need to know the hysteresis—differential hysteresis among soils?
Is it biologically important?

A. Yes, I think it's biologically important. I do feel that desorption, of
course, is in effect measuring some of this. It's not measuring the total
hysteresis, but it's measuring one leg of it and it is the one with which
we are concerned when modelling or evaluating nutrient availability.
In these two instances and from soils that we have studied, there is a
direct relationship between the type of pattern that we have seen in
the adsorption-desorption curves and the effect that you get in the
buffer power evaluation.

Q. If we could characterize this hysteresis for phosphorus, do you think
it would relate to our past concepts of "phosphorus fixation"?

A. I think it would relate very closely to it, yes. I'm not ready to say that
it would be identical.

Comment

The question on whether hysteresis is important in my view depends
upon the concentration that you're going to have at the root surface,
because you've got ions diffusing from the soil towards the root surface. If that concentration is low enough, so you're really beyond
where this hysteresis affect is occurring, then you're using an average
buffer power, and it really doesn't make much difference whether
you're using an adsorption or desorption isotherm. On the other
hand, if your concentration at the root surface is a lot higher, then it
will make a difference. In most cases the concentration at the root is
extremely low. On another point that came up, I don't think we
should be worried about equilibrium conditions because we don't
have equilibrium when plants adsorb the nutrients in the soil.

Comment

I'd like to make this comment, too. I did not intend to raise a question
about equilibrium. I was raising a question about the saturation level;
that one is concerned with adsorption compared to desorption. In
other words, are different sites involved during adsorption compared
to desorption. It's true that there are two different curves and obviously there is an explanation for the two; we ought to be looking at
the explanation and not at the fact that we have it.

Comment

I agree with your point. I think this was illustrated by the Zanesville
soil, since it seemed to have a group of sites that were very retentive
at the low concentrations compared to other soils that we have
studied. These sites appeared to behave differently during adsorption
as compared to desorption.

Comment

Dr. Barber does make a good point as have all of you. When you're
talking about the bioavailability of an element, if you exceed the level
where you're approaching maximum rates of uptake, it doesn't
matter what you do or what you measure because you won't measure

a biological response anyway. But as you move down into the critical range, where an increment of increased activity at the root surface brings on an increase of bioavilability, it's very important. It also is the low concentrations, and not the high ones where many important reactions occur. I think everybody would agree with that. So we don't want to lose sight of the fact that one of the objectives of this is to discuss what are the critical levels and how do you express the bioavailability of them at the critical level.